Introduction to Conservation of Wildlife in India

By Siva Prasad Bose

Published by Joy Bose

Contents

Glossary of Key Terms

Biodiversity: The variety of life on Earth at all its levels, from genes to ecosystems, encompassing the evolutionary, ecological, and cultural processes that sustain life.

Biosphere Reserve: A UNESCO-recognized area that combines biodiversity conservation with sustainable development, typically consisting of a core protected zone, a buffer zone, and a surrounding transition area.

Carrying Capacity: The maximum number of individuals of a species that a habitat can support indefinitely, given the available food, water, shelter, and other resources.

CITES: Convention on International Trade in Endangered Species of Wild Fauna and Flora. An international agreement between governments to ensure that international trade in specimens of wild animals and plants does not threaten their survival.

Conservation: The management of the natural environment and its resources to maintain ecological health, protect biodiversity, and sustain the processes that support life, while allowing for the intelligent use of natural resources by present and future generations.

Ecological Succession: The process by which the structure of a biological community evolves over time. It involves a series of stages from pioneer species colonizing a bare environment to the eventual establishment of a stable climax community.

Ecosystem: A biological community of interacting organisms and their physical environment. In an ecosystem, plants, animals, and other organisms interact with each other and with the abiotic (non-living) components such as soil, water, and air.

Endemic Species: A species that is native to and found only in a particular geographic region, and is not naturally found elsewhere in the world.

Flagship Species: A species selected as an ambassador or symbol for a defined habitat, issue, campaign, or environmental cause. Flagship species are often large, charismatic animals whose conservation efforts also benefit many other species sharing their habitat.

Habitat Fragmentation: The process by which a large, continuous habitat is divided into smaller, isolated patches. This is typically caused by human activities such as deforestation, urbanization, and infrastructure development, and poses a major threat to wildlife populations.

IUCN Red List: The International Union for Conservation of Nature's comprehensive inventory of the global conservation status of plant and animal species. Categories include Extinct, Critically Endangered, Endangered, Vulnerable, Near Threatened, and Least Concern.

Metapopulation: A group of spatially separated populations of the same species that interact at some level through immigration and emigration. Metapopulation theory is critical to conservation planning for species living in fragmented landscapes.

National Park: A protected area reserved for conservation purposes and typically maintained by a national government. In India, national parks have the highest level of protection, with human settlement, agriculture, and most economic activities strictly prohibited within their boundaries.

Poaching: The illegal hunting, capture, or collection of wildlife, including plants and animals, in violation of local or international law. Poaching is one of the most serious threats to biodiversity and pushes many species towards extinction.

Preservation: The protection of nature from human use and development, allowing natural processes to operate without interference. Unlike conservation, preservation does not involve active management; it focuses on protecting areas and species from any form of exploitation.

Trophic Cascade: An ecological phenomenon triggered by the addition or removal of a top predator and resulting in changes in population sizes and behaviors through the food web. The loss of apex predators like tigers can cause dramatic changes throughout an ecosystem.

Wildlife Corridor: A strip of habitat connecting wildlife populations separated by human activities or structures. Corridors allow animals to move between habitat patches, enabling breeding, foraging, and migration, thereby maintaining genetic diversity and population viability.

Wildlife Sanctuary: A protected area in India where wildlife and their habitats are protected, but limited human activities such as controlled grazing and regulated tourism are permitted. Wildlife sanctuaries often serve as buffer zones around national parks or as standalone reserves for specific species.

Glossary of Key Terms

About the author

Other Books by Siva Prasad Bose

Dedication

This book is dedicated to all those who have been involved in conservation of the wildlife, ecology and biodiversity of India, including all kinds of animal and plant life.

Preface

India has a rich variety of biodiversity, including many national parks and nature reserves. Sadly, much of its biodiversity is in danger due to many factors including a fast-growing population and an over emphasis on development. Forest land and agricultural land are often cut to make way for housing. There needs to be a concerted effort to save India's ecology and wildlife.

In this book we introduce to the reader some concepts and issues related to wildlife conservation in India. We discuss important wildlife species, laws, protected areas and sanctuaries and organizations in the field of wildlife conservation in India.

It is hoped that this book will provide the interested reader with useful information about conservation of natural wildlife. Each chapter is written to be accessible to the general reader while maintaining accuracy and depth. The book covers both the ecological principles underpinning conservation science and the practical legal, institutional, and community-based mechanisms that India has developed to protect its wildlife.

Acknowledgements

In preparing this book, the authors would like to acknowledge help from the following sources:

- Conservation and Management of Biological Resources in Himalaya. Editors PS Ramakrishnan, AN Purohit, KG Saxena, KS Rao, RK Maikhuri. Oxford and IBH Publishing Company Co Pvt. Ltd. 1996.

- Indian Institute of Ecology and Environment. Occasional Monographs from the Ecology Course. 1992.

- Conservation Kaleidoscope: People, Protected Areas and Wildlife in Contemporary India by Pankaj Sekhsaria. Kalpavriksh. 2021

- Ecology, Third Edition. Begon, Harper, Townsend. Blackwell Science. 1996

- The Ecology Book: Big Ideas Simply Explained. DK. Tony Juniper. 2019.

- Indian Wildlife. By M.K. Ranjitsinh. Brijbasi. 1995

- Wildlife Conservation and Management by Reena Mathur. Rastogi Publications. 2019

Chapter 1: Introduction to Wildlife Conservation

In this chapter, we discuss some of the principles and issues involved in the area of wildlife conservation.

1.1 Conservation vs Preservation of Wildlife

We first discuss two similar but slightly different terms: conservation and preservation of wildlife. These terms have a different meaning for wildlife than when used in the context of conservation and preservation of buildings.

Both these terms, conservation and preservation are often used when talking of wildlife, but they have slightly different meanings:

Conservation is where humans have a direct and active role to play to keep the balance between wildlife and the environment.

It involves the following:

- Knowing the carrying capacity of a particular area (capacity of the area to support all kinds of life as per the resources available) where the wildlife is residing.

- Conservation focuses on the intelligent use of natural resources, limiting the number of animals as per the carrying capacity of an area and monitoring them so that they can be useful to present and future generations.

For example, conservation may allow controlled hunting to keep the animal population under control.

Preservation, on the other hand, refers to the idea that wildlife should be left alone and not be exploited or used at all.

This includes the following:

- Preservation is not about managing wildlife in an area intelligently, but simply preserving it and let nature do the managing.

- Unlike conservation, carrying capacity of an area is not relevant to preservation because the area is to be left untouched, so its carrying capacity would not be changed.

- Preservation can be achieved by creation of sanctuaries, wildlife reserves and national parks.

Thus, both conservation and preservation are useful and good for wildlife, but each of these should be used intelligently and differently for different animal species.

1.2 Meaning of Wildlife Conservation

Wildlife conservation is a critical field dedicated to the protection and management of the world's diverse species and their habitats. It encompasses a range of practices and strategies aimed at preserving wildlife populations, maintaining ecological balance, and ensuring that natural environments continue to thrive. This chapter provides an overview of wildlife conservation, highlighting its importance and the main issues, with a particular focus on challenges faced in regions like India.

Wildlife conservation is essential for a variety of reasons including the following:

- **Biodiversity Preservation**: Conserving wildlife helps maintain biodiversity, which is vital for ecosystem health and stability. Diverse ecosystems support a wide range of species, each playing a unique role in ecological processes.

- **Ecosystem Services**: Healthy wildlife populations contribute to ecosystem services such as pollination, seed dispersal, and soil formation. These services are crucial for agricultural productivity, water quality, and climate regulation.

- **Cultural and Economic Value**: Wildlife has significant cultural, spiritual, and economic value. Many communities depend on wildlife for their livelihoods, and species like tigers and elephants are integral to cultural heritage and tourism industries.

1.3 Challenges in Wildlife Conservation in India

Wildlife conservation faces several pressing challenges, particularly in biodiversity-rich regions like India:

Habitat Loss and Degradation: The expansion of agriculture, urbanization, and infrastructure development leads to habitat loss and fragmentation. In India, rapid development has encroached upon critical wildlife habitats, threatening species such as the Bengal tiger and the Indian elephant. Habitat fragmentation also disrupts wildlife corridors, impeding species migration and breeding.

Poaching and Illegal Wildlife Trade: Poaching and illegal wildlife trade pose severe threats to many species. Poaching for ivory, rhino horns, and other animal parts drives many species to the brink of extinction. In India, poaching networks target iconic species like tigers and rhinos, driven by high demand for wildlife products in illegal markets.

Human-Wildlife Conflict: As human populations expand into wildlife habitats, conflicts between humans and animals become more frequent. Incidents such as crop raiding by elephants or attacks by predators result in economic losses and human casualties. Managing these conflicts is crucial for the coexistence of wildlife and human communities.

Climate Change: Climate change impacts wildlife through altered weather patterns, temperature changes, and shifting habitats. In India, changing monsoon patterns and rising temperatures affect species distribution and ecosystem dynamics. For example, shifting temperatures in the Himalayas impact the snow leopard's habitat and prey availability.

Pollution: Pollution from agricultural runoff, industrial waste, and plastic pollution affects wildlife health and habitats. Contaminants in water sources and soils can poison wildlife and disrupt reproductive processes. In India, pollution threatens aquatic species in rivers and coastal areas, impacting biodiversity and ecosystem health.

1.4 Conclusion

In this chapter, we have discussed the meaning and challenges in the field of wildlife conservation.

Chapter 2: Wildlife Conservation in the Context of Wider Conservation

Wildlife conservation is a crucial component of broader environmental conservation, which encompasses various branches aimed at maintaining ecological balance and ensuring sustainable use of natural resources. Understanding wildlife conservation within this wider context highlights its interconnections with other environmental efforts and underscores its importance in global sustainability.

Figure: An illustration of ecosystem conservation

2.1. Ecosystem Conservation

Wildlife conservation is deeply intertwined with ecosystem conservation, which focuses on protecting entire ecosystems to ensure their health and functionality. Healthy ecosystems provide essential services such as clean air and water, soil fertility, and climate regulation, which are vital for the survival of wildlife. For example, preserving tropical rainforests not only protects species like jaguars and orangutans but also maintains the broader ecological processes that support diverse life forms.

Figure: An illustration of habitat restoration

2.2 Habitat Restoration

Habitat restoration is a key branch of conservation that aims to rehabilitate degraded environments to their natural state. Restoring habitats such as wetlands, forests, and grasslands supports wildlife by providing necessary resources and conditions for their survival. Initiatives like reforestation and wetland rehabilitation enhance the availability of food, shelter, and breeding grounds for species, contributing to their recovery and resilience.

2.3 Biodiversity Conservation

Biodiversity conservation encompasses efforts to protect the variety of life forms on Earth, including species, ecosystems, and genetic diversity. Wildlife conservation is a fundamental aspect of biodiversity conservation, as it focuses on preserving animal species and their habitats. Efforts to conserve biodiversity also involve protecting plant species, which are crucial for maintaining ecosystem health and supporting wildlife.

2.4 Wildlife conservation and Climate Change Mitigation

Wildlife conservation and climate change mitigation are closely linked. Climate change affects wildlife through habitat loss, altered food sources, and shifting migration patterns. Conservation strategies that address climate change—such as creating climate-resilient protected areas and reducing greenhouse gas emissions—also benefit wildlife by helping them adapt to changing conditions and maintain stable habitats.

2.5 Wildlife Conservation and Sustainable Development

Sustainable development integrates environmental conservation with economic and social goals. Wildlife conservation supports sustainable development by promoting practices that balance human needs with environmental protection. Initiatives such as eco-tourism and sustainable agriculture provide economic benefits while conserving wildlife and their habitats. Engaging local communities in conservation efforts ensures that development is inclusive and benefits both people and wildlife.

2.6 Wildlife Conservation and Pollution Control

Addressing pollution is essential for wildlife conservation. Pollution from industrial activities, agriculture, and urban areas can harm wildlife through habitat degradation, contamination of food sources, and health issues. Conservation efforts that include pollution control measures, such as reducing plastic waste and managing chemical runoff, help protect wildlife and improve ecosystem health.

2.7 Conclusion

Wildlife conservation is a vital element of the broader conservation framework, which includes ecosystem conservation, habitat restoration, and biodiversity protection. By understanding and addressing the interconnectedness of wildlife conservation with other environmental efforts, we can develop more holistic and effective strategies for preserving our planet's natural heritage.

Chapter 3: Principles of Ecology Applied to India's Wildlife Conservation

Ecology is the study of how living organisms interact with each other and their environment. In the context of wildlife conservation, understanding these ecological principles is crucial for the development of effective strategies to protect species and their habitats. India, with its diverse ecosystems ranging from tropical forests to deserts and coastal regions, offers unique challenges and opportunities for applying these principles. This chapter explores key ecological concepts and their relevance to India's wildlife conservation efforts.

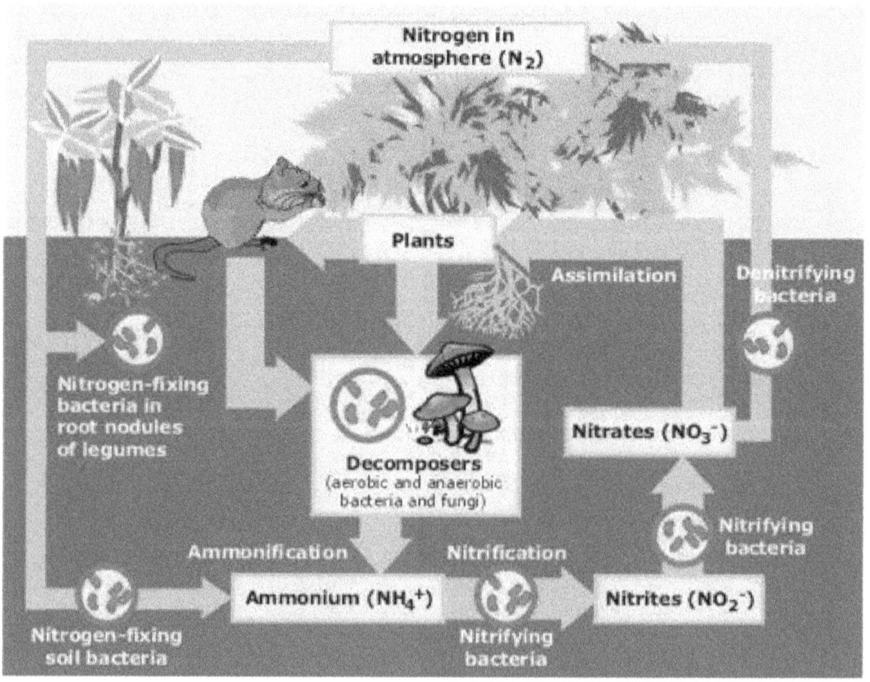

Figure: Nitrogen cycling in an ecosystem. U.S. Environmental Protection Agency, Public domain, via Wikimedia Commons

3.1 Ecosystems and Biodiversity

An ecosystem comprises all living organisms (plants, animals, and microorganisms) in a given area, interacting with each other and their physical environment. India is home to several types of ecosystems, including forests, grasslands, wetlands, mountains, and marine environments. Each ecosystem supports a unique assemblage of species, contributing to India's rich biodiversity.

Biodiversity forms the backbone of ecological stability. It ensures that ecosystems are resilient and capable of recovering from disturbances such as natural disasters, climate change, or human activities. In India, protecting biodiversity is key to sustaining ecosystems like the Western Ghats rainforests, the Sundarbans mangroves, and the Himalayan alpine meadows. The conservation of these ecosystems is crucial for maintaining ecological services such as water regulation, soil fertility, and climate stabilization.

The more diverse an ecosystem, the more resilient it is to change. Conservation strategies in India prioritize biodiversity hotspots, such as the Western Ghats and Eastern Himalayas, as these areas are critical to maintaining the ecological balance and long-term survival of species.

3.2 Habitat and Niche

A species' habitat refers to the physical environment in which it lives, while its niche represents the role it plays within that environment, including its interactions with other species and its use of resources. Every species has a unique niche that defines how it obtains food, reproduces, and avoids predators.

In India, habitat degradation and fragmentation are major threats to wildlife. As human populations grow and development expands, many species are forced out of their natural habitats. For example, tigers and elephants, once widespread across the country, are now confined to fragmented habitats in protected reserves.

Conservation programs such as the creation of wildlife corridors between fragmented habitats are critical to allowing species like elephants to migrate and maintain their ecological niches. Protecting habitats is essential for preserving not just individual species but also the complex web of interactions they rely on.

Figure: Illustration of a wildlife food chain in India

3.3 Food Chains and Trophic Levels

An ecosystem's food chain consists of different trophic levels: producers (plants), primary consumers (herbivores), secondary consumers (carnivores), and decomposers. The energy transfer between these levels is an essential part of ecosystem functioning. Apex predators, like tigers and leopards, occupy the highest trophic levels and play a crucial role in controlling herbivore populations, thus maintaining the balance of the ecosystem.

The loss of top predators can lead to a phenomenon known as trophic cascade, where the absence of a predator allows herbivore populations to explode, which in turn can lead to overgrazing and degradation of vegetation. In India, the decline in predator populations due to poaching or habitat loss can have severe consequences for ecosystems.

Programs like Project Tiger and Project Snow Leopard focus on conserving apex predators, recognizing their role in maintaining ecosystem health. Protecting these species ensures that entire food chains remain balanced, benefiting numerous other species in the ecosystem.

3.4 Carrying Capacity

The carrying capacity of an ecosystem is the maximum number of individuals of a species that it can support without degrading the habitat. If a population exceeds its carrying capacity, it can lead to overexploitation of resources, habitat degradation, and an increased risk of conflict with humans.

In many parts of India, human-wildlife conflict arises when species such as elephants or leopards exceed the carrying capacity of their shrinking habitats, forcing them to venture into human-dominated landscapes in search of food and space.

Wildlife managers must monitor populations and ensure that species numbers remain in balance with the available resources. Efforts to restore degraded habitats, as well as create buffer zones and protected corridors, help manage carrying capacity and reduce conflict.

Figure: An image depicting ecological succession, from barren land to a forest ecosystem

3.5 Ecological Succession and Habitat Restoration

Ecological succession refers to the process by which ecosystems change over time. When an area is disturbed—whether by natural events such as floods or by human activities like deforestation—new species may colonize the area, leading to gradual changes in the ecosystem.

In India, many conservation efforts focus on habitat restoration, which involves rehabilitating degraded ecosystems to restore them to their natural state. Examples include reforesting degraded lands in the Western Ghats and restoring mangrove forests along the coastlines of Gujarat and Odisha.

Understanding the stages of ecological succession is crucial for designing effective restoration projects. Conservation efforts in India must promote native species and natural regeneration processes to restore ecosystems and create habitats that support wildlife in the long term.

3.6 Population Dynamics and Metapopulation Theory

Population dynamics involves studying how populations change in size, density, and structure over time, influenced by factors such as birth rates, death rates, immigration, and emigration. In fragmented landscapes, the metapopulation theory becomes relevant. A metapopulation is a group of spatially separated populations that interact through migration. For species living in fragmented habitats, such as tigers in isolated reserves, the exchange of individuals between populations is essential for maintaining genetic diversity and population viability.

Conservation programs in India must ensure connectivity between fragmented habitats to facilitate the movement of species between populations. Initiatives like the Elephant Corridor Project and the creation of tiger corridors between reserves are based on the principles of metapopulation dynamics, ensuring long-term survival through genetic exchange.

3.7 Human-Wildlife Interactions and Coexistence

Human-wildlife interactions are central to ecology, especially in a country like India, where many people live in close proximity to wildlife. Understanding the behavior of species in response to human activities and developing strategies for coexistence is critical to ensuring both human and wildlife well-being.

In areas like the Western Ghats and Assam, where human settlements overlap with wildlife habitats, conflict arises when animals damage crops, livestock, or even human life. Conservationists focus on creating buffer zones, improving awareness, and compensating communities for losses to foster coexistence.

Programs that promote coexistence, such as compensatory schemes for farmers affected by wildlife and the installation of barriers or warning systems, are essential in managing human-wildlife interactions. Ecologically sound approaches like rewilding and promoting natural prey species for predators help minimize conflict.

3.8 Conclusion

The application of ecological principles is fundamental to wildlife conservation in India. By understanding how ecosystems function, how species interact within them, and how human activities impact these dynamics, conservationists can design more effective strategies to protect the country's rich biodiversity. From maintaining predator populations to restoring degraded habitats and ensuring the connectivity of fragmented landscapes, ecological science is at the heart of every successful conservation effort in India.

Chapter 4: International Treaties and Conventions Related to Wildlife Conservation

International treaties and conventions play a crucial role in coordinating global efforts to protect wildlife and their habitats. These agreements set standards, promote cooperation among nations, and provide frameworks for addressing transboundary environmental issues. For India, participating in these international agreements is vital for enhancing its wildlife conservation strategies and addressing global challenges.

4.1. Convention on International Trade in Endangered Species of Wild Fauna and Flora (CITES)

Established in 1973, CITES is a global agreement aimed at ensuring that international trade in specimens of wild animals and plants does not threaten their survival. It classifies species into three appendices based on their level of threat and regulates trade accordingly. India, as a party to

CITES, has implemented measures to control the trade of endangered species like tigers and rhinos, reducing illegal wildlife trade and protecting critically endangered species.

4.2. Convention on Biological Diversity (CBD)

The CBD, adopted in 1992 at the Earth Summit in Rio de Janeiro, focuses on the conservation of biodiversity, sustainable use of its components, and fair sharing of benefits arising from genetic resources. India's commitment to the CBD includes developing national biodiversity strategies and action plans, creating protected areas, and engaging in global efforts to conserve ecosystems and species.

4.3. Ramsar Convention on Wetlands

The Ramsar Convention, established in 1971, aims to conserve wetlands of international importance. Wetlands are crucial habitats for a variety of wildlife, including migratory birds and aquatic species. India has designated several Ramsar sites, such as the Keoladeo National Park and Chilika Lake, to protect these vital ecosystems and their biodiversity.

4.4. Convention on Migratory Species (CMS)

The CMS, also known as the Bonn Convention, addresses the conservation of migratory species and their habitats. It promotes international cooperation to protect migratory species throughout their ranges. India is involved in CMS agreements to safeguard migratory species such as the Amur falcon and various marine turtles that migrate through Indian waters.

4.5. UNESCO World Heritage Convention

The World Heritage Convention, adopted in 1972, aims to protect cultural and natural heritage sites of outstanding universal value. Several of India's national parks and wildlife sanctuaries, including Kaziranga National Park and the Western Ghats, are recognized as World Heritage Sites due to their exceptional biodiversity and ecological significance.

4.6. Global Conservation Initiatives

Global wildlife conservation efforts are characterized by collaborative projects, international funding mechanisms, and multinational partnerships. Key initiatives include:

Global Environment Facility (GEF): The GEF provides financial support for projects aimed at addressing global environmental challenges, including biodiversity conservation. Indian projects funded by the GEF have focused on ecosystem restoration, wildlife protection, and sustainable land management.

International Union for Conservation of Nature (IUCN): The IUCN works globally to conserve nature and ensure sustainable use of natural resources. Its Red List of Threatened Species is a valuable resource for identifying and prioritizing conservation actions for endangered species.

Wildlife Conservation International (WCI): WCI supports conservation projects worldwide, including in India, to protect endangered species and their habitats through research, advocacy, and community engagement.

4.6 Benefits to India of global wildlife conservation efforts

Participating in and benefiting from global wildlife conservation efforts provides several advantages for India:

Enhanced Funding and Resources: International treaties and global conservation organizations offer financial support, technical expertise, and resources for wildlife conservation projects. This support helps India implement effective conservation strategies and address challenges such as habitat loss and poaching.

Knowledge Sharing and Best Practices: Global collaborations facilitate the exchange of knowledge, research findings, and best practices in wildlife conservation. India benefits from international expertise in areas such as species monitoring, habitat management, and anti-poaching technologies.

Strengthened Conservation Networks: International treaties and organizations foster partnerships and networks that enhance conservation efforts. India's involvement in global initiatives strengthens its connections with other countries and conservation groups, leading to coordinated efforts and increased effectiveness.

Policy and Legal Frameworks: International agreements influence national policies and legal frameworks, guiding India's conservation priorities and actions. Compliance with global standards ensures that India's conservation measures align with international best practices and contribute to global biodiversity goals.

4.7 Conclusion

International treaties and conventions play a pivotal role in wildlife conservation by setting standards, promoting cooperation, and providing resources and frameworks for addressing global environmental challenges. For India, engaging with these global efforts enhances its capacity to protect its rich biodiversity and address pressing conservation issues. By leveraging international support and participating in collaborative initiatives, India can strengthen its wildlife conservation strategies and contribute to global efforts to preserve the planet's natural heritage.

Chapter 5: Threats to India's Wildlife

India's diverse ecosystems, ranging from dense rainforests and grasslands to deserts and coastal habitats, are home to an incredible variety of wildlife. From the majestic Bengal tiger and the Asiatic elephant to the elusive snow leopard and the endangered gharial, India's fauna is globally significant. However, this biodiversity faces a range of escalating threats that jeopardize the survival of many species and ecosystems. This chapter explores the major threats to India's wildlife, examining their causes and potential impacts on the country's rich natural heritage.

5.1 Habitat Loss and Fragmentation

Habitat loss is perhaps the most significant threat to wildlife in India. As human populations grow and expand, forests and grasslands are being cleared for agriculture, urbanization, industrial development, and infrastructure projects.

The causes include the following:

Deforestation: Forests are cleared for timber, agricultural expansion, and infrastructure projects like roads and dams. This not only destroys the habitat of countless species but also reduces biodiversity and ecological stability.

Agricultural Expansion: With over 60% of India's population dependent on agriculture, more land is converted into farmland every year, encroaching on wildlife habitats.

Urbanization: The expansion of cities and towns into previously undeveloped areas puts tremendous pressure on surrounding ecosystems, further fragmenting habitats.

Habitat fragmentation isolates wildlife populations, preventing them from moving freely between areas, which is crucial for foraging, breeding, and genetic diversity. For example, tigers and leopards in many parts of India are now confined to small, disconnected reserves, which reduces their chance of long-term survival. Habitat fragmentation also increases the likelihood of human-wildlife conflicts.

Figure: Illustration of poaching of rhinos in India

5.2 Poaching and Illegal Wildlife Trade

Poaching and illegal wildlife trade pose an immediate and direct threat to India's wildlife. Many species are hunted for their body parts, which are sold in both domestic and international markets.

Causes of poaching include the following:

Demand for Wildlife Products: The illegal trade in wildlife products, including tiger skins, elephant ivory, rhino horns, pangolin scales, and bear bile, is driven by high demand in traditional medicine, luxury markets, and even the pet trade.

Lack of Enforcement: While India has strong wildlife protection laws, enforcement remains a challenge, particularly in remote areas where illegal activities are harder to monitor.

Several species are on the brink of extinction due to poaching. The tiger population, for instance, has drastically declined over the past century due to poaching for their skins and bones. Similarly, poaching continues to push species like the one-horned rhinoceros, pangolins, and various species of turtles and birds toward extinction.

5.3 Human-Wildlife Conflict

As human populations expand into wildlife habitats, conflicts between humans and wildlife have become more frequent. These conflicts occur when wildlife, searching for food or territory, come into contact with human settlements, often resulting in harm to both humans and animals.

Causes of human wildlife conflict include the following:

Encroachment on Habitats: Expanding agricultural fields, roads, and settlements near forests and protected areas often lead to wildlife entering human-dominated landscapes.

Depletion of Natural Prey: In some cases, overhunting or habitat degradation reduces the availability of natural prey for carnivores, forcing them to attack livestock or, in rare cases, humans.

Elephants frequently raid crops, while leopards and tigers sometimes prey on livestock, leading to economic losses for farmers. This often results in retaliatory killings of wildlife. The human cost can also be high, with incidents of people being injured or killed during such encounters. These conflicts create a negative perception of wildlife among local communities, complicating conservation efforts.

5.4. Climate Change

Climate change is an emerging and long-term threat to India's wildlife. Altered weather patterns, rising temperatures, and changing precipitation rates have a profound impact on ecosystems and species distribution.

The causes of climate change include:

Temperature Rise: Global warming leads to habitat shifts, with species needing to migrate to higher altitudes or latitudes to survive. For instance, species in the Western Ghats and Himalayas are particularly vulnerable.

Changes in Precipitation: Erratic monsoons and prolonged droughts can alter freshwater availability and disrupt seasonal patterns of breeding, migration, and food supply.

Species with narrow ranges, such as the snow leopard and other alpine wildlife, are at particular risk as their habitats shrink. Additionally, marine species like coral reefs and mangroves, which serve as important nurseries for marine life, are highly sensitive to ocean temperature rise and acidification, leading to biodiversity loss in India's coastal and marine ecosystems.

5.5 Invasive Species

Invasive species are non-native plants, animals, or pathogens that spread into new ecosystems and outcompete native species. In India, invasive species pose a significant threat to biodiversity and ecosystem health.

The causes of invasive species include:

Human Activities: Invasive species are often introduced through trade, travel, or agriculture. Once established, they can spread rapidly, especially in disturbed ecosystems.

Lack of Natural Predators: Invasive species often have no natural predators in their new environments, allowing them to proliferate unchecked.

Invasive plant species such as Lantana camara and Prosopis juliflora have spread widely in Indian forests, outcompeting native vegetation and reducing food availability for herbivores. Invasive fauna like the African catfish and house crows disrupt aquatic and avian ecosystems, respectively, threatening native species and altering ecosystem dynamics.

5.6. Pollution

Pollution, in its many forms—air, water, and soil pollution—has detrimental effects on wildlife and their habitats.

Causes of pollution include:

Water Pollution: Rivers and wetlands are increasingly polluted by industrial waste, agricultural runoff, and untreated sewage. Toxic substances like heavy metals and pesticides can accumulate in aquatic ecosystems, affecting both flora and fauna.

Plastic Waste: The widespread use of plastic has resulted in plastic pollution in terrestrial and marine environments, harming wildlife through ingestion or entanglement.

Air Pollution: Air pollution, especially in industrial areas and urban centers, affects wildlife health. It is particularly problematic for bird species that depend on clean air and aquatic ecosystems.

In India's rivers, species like the Ganges River dolphin and gharial are highly sensitive to water pollution, which contaminates their habitats and disrupts their food supply. Similarly, plastic pollution in marine environments poses severe risks to marine species like sea turtles and seabirds.

5.7. Overexploitation of Resources

Overexploitation of natural resources, such as overgrazing by livestock, unsustainable fishing practices, and excessive extraction of forest products, poses a significant threat to India's wildlife.

Causes of over-exploitation include the following:

Overfishing: Unsustainable fishing practices deplete marine species, threatening India's marine biodiversity.

Livestock Grazing: Overgrazing in forests and grasslands depletes the natural vegetation that herbivores like deer and antelope rely on, indirectly affecting the predators that feed on them.

Overexploitation reduces the carrying capacity of ecosystems, leading to degraded habitats and dwindling wildlife populations. For example, unsustainable fishing practices in India's coastal waters threaten species such as sea turtles, dolphins, and sharks, as well as coral reef ecosystems.

5.8 Conclusion

India's wildlife faces a multitude of threats, driven largely by human activities and exacerbated by environmental changes. Habitat loss, poaching, climate change, and pollution continue to push many species toward extinction, while human-wildlife conflict and invasive species further complicate conservation efforts.

Chapter 6: Biodiversity and Sustainable Development in India and especially Himalayas

Biodiversity and sustainable development are deeply interconnected concepts. Biodiversity, the variety of life in all its forms, from ecosystems to species, plays a critical role in maintaining ecosystem services that support human well-being. In India, where ecosystems range from the arid deserts of Rajasthan to the lush forests of the Western Ghats and the icy heights of the Himalayas, managing biodiversity sustainably is crucial to the country's economic, social, and environmental stability. This chapter explores the relationship between biodiversity and sustainable development in India, with a particular focus on the management of biodiversity in the Himalayan region.

6.1 Importance of Biodiversity for Sustainable Development

Biodiversity is essential for the functioning of ecosystems, which in turn support human livelihoods. In India, millions of people depend directly on ecosystems for their daily needs—food, water, medicine, and fuel—especially in rural and indigenous communities. Biodiversity also supports agriculture, fisheries, forestry, and tourism, contributing significantly to the economy.

Key ecosystem services Supported by Biodiversity include:

Provisioning Services: Resources such as timber, fresh water, food (crops, fish), and medicinal plants come from biodiverse ecosystems.

Regulating Services: Ecosystems help regulate the climate, control floods, purify water, and sequester carbon.

Cultural Services: Biodiversity contributes to spiritual, recreational, and educational values, which are particularly important in India's cultural and religious contexts.

Supporting Services: Biodiversity underpins essential services like soil formation, nutrient cycling, and pollination, which are vital for agriculture and ecosystem productivity.

Sustainable development, as defined by the United Nations, seeks to meet the needs of the present without compromising the ability of future generations to meet their own needs. In India, maintaining biodiversity is fundamental to achieving sustainable development goals (SDGs), particularly those related to poverty eradication, food security, climate action, and life on land and below water.

6.2 Challenges to Biodiversity in India

India is one of the world's megadiverse countries, home to about 8% of global species diversity. However, this biodiversity is under serious threat from various factors such as habitat loss, climate change, overexploitation, pollution and invasive species as mentioned in the previous chapters.

To address these challenges, India must strike a balance between economic development and biodiversity conservation, particularly in ecologically sensitive regions like the Himalayas.

6.3 Biodiversity in the Himalayan Region

The Indian Himalayan region, stretching across the northern states of Jammu & Kashmir, Himachal Pradesh, Uttarakhand, Sikkim, and Arunachal Pradesh, is a globally significant biodiversity hotspot. The Himalayas are home to a wide range of ecosystems, from subtropical forests in the foothills to alpine meadows and glaciers at higher altitudes. The region supports an extraordinary array of species, including iconic animals like the snow leopard, red panda, Himalayan musk deer, and a rich diversity of flora and fauna.

6.4 Ecological importance of Himalayas

Ecological Importance of the Himalayas includes the following:

Water Towers of Asia: The Himalayas feed major rivers like the Ganges, Brahmaputra, and Indus, providing water to hundreds of millions of people across India.

Carbon Sequestration: Forests in the Himalayas play a crucial role in absorbing carbon dioxide, helping to mitigate climate change.

Biodiversity Reservoir: The region harbors numerous endemic species, many of which are not found anywhere else in the world.

6.5 Human Dependency in Himalayan communities

The Himalayan communities rely on natural resources for their livelihoods. Agriculture, forestry, livestock grazing, and tourism are the mainstay of the local economy. However, unsustainable land use, deforestation, and overgrazing are degrading the fragile ecosystems of the region.

6.6 Sustainable Management of Himalayan Biodiversity

Sustainable development in the Himalayas must balance conservation goals with the economic and social needs of local communities. Several strategies are being implemented to achieve this:

Integrated Ecosystem Management: This approach focuses on managing entire ecosystems, rather than individual species, to ensure long-term ecological stability. Programs like the **National Mission for Sustaining the Himalayan Ecosystem (NMSHE)** promote sustainable management of natural resources, conservation of glaciers, and restoration of degraded ecosystems in the Himalayan region.

Community-Based Conservation: Local communities play a central role in the conservation of biodiversity in the Himalayas. By involving indigenous people in conservation efforts, such as Community Conserved Areas (CCAs), the government and NGOs aim to promote sustainable land use practices, protect traditional knowledge, and ensure local ownership of conservation initiatives. For example, in Uttarakhand, local villagers are involved in forest management through Van Panchayats, community forest councils that promote sustainable forest use.

Promotion of Ecotourism: Ecotourism is an important component of sustainable development in the Himalayas. It provides an alternative income source for local communities while promoting the conservation of wildlife and ecosystems. Popular destinations such as Hemis National Park in Ladakh, known for its snow leopard population, have become models for ecotourism that contributes to both conservation and livelihood development.

Climate Adaptation Strategies: Climate change poses a significant threat to the Himalayan region, particularly due to the melting of glaciers and shifting weather patterns. India's **National Action Plan on Climate Change (NAPCC)** includes initiatives aimed at conserving mountain ecosystems, improving water management, and supporting local communities in adapting to climate impacts. Projects like sustainable agriculture practices, rainwater harvesting, and afforestation are helping communities cope with changing conditions.

Conservation of Flagship Species: Conserving flagship species like the snow leopard is key to protecting broader ecosystems. The Snow Leopard Project works to protect this apex predator, whose conservation helps maintain the balance of the entire alpine ecosystem. Similarly, the conservation of the Himalayan Monal, the state bird of Uttarakhand, ensures the protection of high-altitude grasslands.

Legal and Policy Frameworks: India has a robust legal and policy framework to protect its biodiversity and promote sustainable development. These include **Wildlife Protection Act** (1972), which provides the foundation for wildlife conservation and protected areas in India. **The Biodiversity Act** (2002) aims to conserve biological diversity, promote its sustainable use, and ensure the equitable sharing of benefits arising from the use of genetic resources. The **National Biodiversity Authority (NBA)** was established to implement the Biodiversity Act, the NBA plays a crucial role in regulating the use of biological resources and ensuring the participation of local communities in conservation efforts.

International Commitments: India is also a signatory to international agreements such as the **Convention on Biological Diversity (CBD)** and the Paris Agreement on Climate Change, reinforcing its commitment to biodiversity conservation and sustainable development.

6.7 Conclusion

Biodiversity and sustainable development are closely linked in India, particularly in ecologically sensitive regions like the Himalayas. The country's approach to biodiversity management, involving integrated ecosystem management, community participation, and climate adaptation, offers a pathway to sustainable development that benefits both people and nature. However, ongoing efforts are needed to address the complex challenges posed by habitat loss, human-wildlife conflict, and climate change, ensuring that India's rich biodiversity is preserved for future generations.

Chapter 7: Wildlife Protection Act, Forest Conservation Act and Biodiversity Act

In this chapter we discuss the Wildlife Protection Act 1972, which seeks to protect the plants and animal species of India. We also discuss the Forest Conservation Act 1980.

7.1 Summary of Wildlife Protection Act

After decades of poaching and smuggling and encroaching into forests, wildlife in India has become threatened. Several species such as the royal Bengal tiger are endangered. Illegal poaching of tigers and leopards for their skin and elephants for ivory are just some examples.

To prevent such incidents and protect the wildlife of India, the Wildlife Protection Act was enacted in 1972 by the Indian parliament. It imposes restrictions on hunting and poaching and illegal trade in wildlife, provides protection for certain species of wildlife. It establishes protected areas for wildlife such as sanctuaries, national parks and protected reserves. It establishes bodies to manage various zoos, national parks and reserves. It also constitutes a national board for wildlife comprising the prime minister, various ministers and secretaries, as well as state boards.

The original act has been amended several times to make it more effective, especially with the 2002 Amendment Act which made the punishments more stringent and provides for improved enforcement.

THE WILD LIFE (PROTECTION) ACT, 1972[*]

ACT NO. 53 OF 1972

[*9th September, 1972.*]

[1][An Act to provide for the protection of wild animals, birds and plants and for matters connected therewith or ancillary or incidental thereto with a view to ensuring the ecological and environmental security of the country.]

[2]*   *   *   *   *

CHAPTER I

PRELIMINARY

1. Short title, extent and commencement.—(*1*) This Act may be called the Wild Life (Protection) Act, 1972.

[3][(*2*) It extends to the whole of India except the State of Jammu and Kashmir.]

(*3*) It shall come into force in a State or Union territory to which it extends [4]*** on such date as the Central Government may, by notification, appoint, and different dates may be appointed for different provisions of this Act or for different States or Union territories.

2. Definitions.—In this Act, unless the context otherwise requires,—

[5][(*1*) "animal" includes amphibians, birds, mammals and reptiles and their young, and also includes, in the cases of birds and reptiles, their eggs;]

(*2*) "animal article" means an article made from any captive animal or wild animal, other than vermin, and includes an article or object in which the whole or any part of such animal [6][has been used, and ivory imported into India and an article made therefrom];

[7]*   *   *   *   *

[8][(*4*) "Board" means a State Board for Wild Life constituted under sub-section (1) of section 6;]

(*5*) "captive animal" means any animal, specified in Schedule I, Schedule II, Schedule III or Schedule IV, which is captured or kept or bred in captivity;

[9]*   *   *   *   *

(*7*) "Chief Wild Life Warden' means the person appointed as such under clause (*a*) of sub-section (*1*) of section 4;

[10][(*7A*) "circus" means an establishment, whether stationary or mobile, where animals are kept or used wholly or mainly for the purpose of performing tricks or manoeuvres;]

[11]*   *   *   *   *

[12][(*9*) "Collector" means the chief officer in charge of the revenue administration of a district or any other officer not below the rank of a Deputy Collector as may be appointed by the State Government under section 18B in this behalf;]

(*10*) "commencement of this Act", in relation to—

Figure: Front page of the wildlife protection Act 1972

The Wildlife Protection Act provides for the protection of Indian wildlife including endangered animals and plants and also birds, whose species are mentioned in lists. It extends to the whole of India.

7.2 Schedules of the Wildlife Protection Act

The act has six schedules which protect wildlife in varying degrees as follows:

- Schedules 1 and 2 contain a list of species of animals who are given the highest level of protection and offences under these have the highest penalties. The prescribed imprisonment for offences is 3 years to 7 years along with a fine of minimum Rs 10000 with 25000 for subsequent offences. Schedule 1 includes species such as cheetah, blackbuck, tiger and Indian lion, as well as protected amphibians and reptiles such as crocodiles and pythons.

- Schedule 3 and 4 have lists of wildlife species which are also protected, however the penalties are lower than schedule 1 and 2.

- Schedule 5 contains a list of pests or vermin which include common crows, fruit bats, rats and mice which can be freely hunted.

- Schedule 6 contains a list of endemic plants that are prohibited from cultivation and planting. Such plants include pitcher plant and blue and red vanda.

7.3 Statutory bodies set up under the wildlife protection Act

The Wildlife protection act also sets up a number of important statutory bodies whose responsibility lies in different aspects of wildlife protection.

These bodies are as follows:

- National Board for Wildlife and state wildlife advisory boards

- Central Zoo Authority

- Wildlife Crime Control Bureau

- National Tiger Conservation Authority

7.4 Role of India's Wildlife Protection Act in Conservation

India's Wildlife Protection Act of 1972 is a cornerstone of the country's efforts to conserve its rich biodiversity. The Act was established in response to growing concerns about the rapid decline of wildlife populations and habitat destruction. It provides a comprehensive legal framework aimed at safeguarding wildlife and their habitats, promoting conservation, and regulating human activities that impact wildlife. This subsection explores how the Wildlife Protection Act has significantly contributed to the conservation of wildlife in India.

Legal Framework for Wildlife Conservation: The Wildlife Protection Act provides a robust legal framework for the protection of wildlife through several key provisions:

Establishment of Protected Areas: The Act empowers the government to designate areas as protected, including National Parks, Wildlife Sanctuaries, and Conservation Reserves. These protected areas serve as safe havens for wildlife, preventing habitat destruction and reducing human-wildlife conflicts. Notable examples include Kaziranga National Park, known for its population of the Indian one-horned rhinoceros, Kaziranga benefits from the protections afforded by the Act. Another example is the Periyar Wildlife Sanctuary, which is renowned for its elephants and tigers and preserved under the Act's regulations, ensuring habitat protection and biodiversity conservation.

Regulation of Hunting and Trade: The Act strictly regulates hunting and the trade of wildlife and wildlife products. It prohibits hunting of most species, except for those specified under certain conditions, and regulates the trade of animal parts and derivatives. This has led to a significant reduction in illegal hunting and poaching, contributing to the recovery of several endangered species. An example is in **Tiger Conservation**, where the act's stringent controls on hunting have been instrumental in the recovery of tiger populations under initiatives such as Project Tiger.

Protection of Endangered Species: The Act provides special protection to endangered and threatened species by listing them in Schedule I and II. These species receive the highest level of protection from exploitation and trade. Examples of species listed under these schedules include the Bengal Tiger and Indian elephant. As an endangered species, the Bengal tiger is afforded comprehensive protection under the Act, contributing to its conservation through habitat preservation and anti-poaching measures. Similarly, the Act's protections help safeguard elephant populations and their migratory corridors from encroachment and habitat loss.

7.5 Enforcement and Implementation of the Wildlife Protection Act

Effective implementation and enforcement are critical to the success of the Wildlife Protection Act. Several mechanisms and institutions support this including:

Wildlife Crime Control Bureau (WCCB): Established under the Act, the WCCB is responsible for combating wildlife crime, including poaching and illegal trade. The Bureau coordinates with state authorities and other agencies to enforce the Act's provisions, conduct investigations, and prosecute offenders.

Forest and Wildlife Officials: Forest and wildlife officials play a crucial role in implementing the Act at the ground level. They are responsible for patrolling protected areas, monitoring wildlife populations, and enforcing regulations. Training and capacity-building programs for these officials enhance their effectiveness in conservation efforts.

Public Awareness and Community Involvement: The Act promotes public awareness and community involvement in wildlife conservation. Educational programs and community engagement initiatives help foster a conservation ethic and encourage local participation in protecting wildlife and their habitats.

7.6 Challenges for the Wildlife Protection Act

Challenges include Enforcement Gaps. Despite the Act's provisions, enforcement challenges persist, including inadequate resources, insufficient training, and corruption. Addressing these gaps through increased funding, better training, and stronger accountability measures is essential for effective implementation.

7.7 Forest Conservation Act 1980

The Forest Conservation Act 1980 is intended to ensure the conservation of India's forests and the resources of the forests.

THE FOREST (CONSERVATION) ACT, 1980

ACT NO. 69 OF 1980

[27*th December*, 1980.]

An Act to provide for the conservation of forests and for matters connected therewith or ancillary or incidental thereto.

BE it enacted by Parliament in the Thirty-first Year of the Republic of India as follows:—

1. Short title, extent and commencement.—(*1*) This Act may be called the Forest (Conservation) Act, 1980.

(*2*) It extends to the whole of India except the State of Jammu and Kashmir.

(*3*) It shall be deemed to have come into force on the 25th day of October, 1980.

2. Restriction on the dereservation of forests or use of forest land for non-forest purpose.— Notwithstanding anything contained in any other law for the time being in force in a State, no State Government or other authority shall make, except with the prior approval of the Central Government, any order directing—

(*i*) that any reserved forest (within the meaning of the expression "reserved forest" in any law for the time being in force in that State) or any portion thereof, shall cease to be reserved;

(*ii*) that any forest land or any portion thereof may be used for any non-forest purpose.

Explanation.—For the purposes of this section "non-forest purpose" means the breaking up or clearing of any forest land or portion thereof for any purpose other than reafforestation.

3. Constitution of Advisory Committee.—The Central Government may constitute a Committee consisting of such number of persons as it may deem fit to advise that Government with regard to—

(*i*) the grant of approval under section 2; and

Figure: First page of the Forest Conservation Act 1980

Forests are a unique natural resource and support a variety of animal and plant life. Due to rapid industrialization and also encroaching of forest land for agriculture and grazing purposes, parts of India's forest land were in danger of deforestation. Hence to preserve the forest cover and to protect the forests from getting cleared, the Forest Conservation Act was brought.

The Forest Conservation Act is a rather short act. It mainly restricts the use of reserved forest land for any other purpose, such as by clearing up the forest land. It restricts the de-reserving of reserved forest land, without prior approval of the central government. It constitutes an advisory committee by the central government for purposes of forest conservation. It also declares penalties for contravening the provisions of the act. The power to make rules under this act lies with the central government alone.

7.8 Biodiversity Act 2002

The Biological diversity act or Biodiversity Act 2002 is aimed to preserve the biological diversity in India and the sharing of benefits from biological resources.

The Biological Diversity Act 2002 aims to preserve the biological diversity of India and check biopiracy, such as foreign companies patenting traditional biological resources or remedies. It was inspired by the United Nations Convention on Biological Diversity (CBD) 1992 in which India too participated and which recognizes the rights of countries to their own biological resources.

The act has the following definition of biological diversity:

"biological diversity" means the variability among living organisms from all sources and the ecological complexes of which they are part and includes diversity within species or between species and of eco-systems;

THE BIOLOGICAL DIVERSITY ACT, 2002

ACT NO. 18 OF 2003

[*5th February, 2003.*]

An Act to provide for conservation of biological diversity, sustainable use of its components and fair and equitable sharing of the benefits arising out of the use of biological resources, knowledge and for matters connected therewith or incidental thereto.

WHEREAS India is rich in biological diversity and associated traditional and contemporary knowledge system relating thereto.

AND WHEREAS India is a party to the United Nations Convention on Biological Diversity signed at Rio de Janeiro on the 5th day of June, 1992;

AND WHEREAS the said Convention came into force on the 29th December, 1993;

AND WHEREAS the said Convention reaffirms the sovereign rights of the States over their biological resources;

AND WHEREAS the said Convention has the main objective of conservation of biological diversity, sustainable use of its components and fair and equitable sharing of the benefits arising out of utilisation of genetic resources;

AND WHEREAS it is considered necessary to provide for conservation, sustainable utilisation and equitable sharing of the benefits arising out of utilisation of genetic resources and also to give effect to the said Convention.

BE it enacted by Parliament in the Fifty-third Year of the Republic of India as follows:—

CHAPTER I

PRELIMINARY

1. Short title, extent and commencement.—(*1*) This Act may be called the Biological Diversity Act, 2002.

(*2*) It extends to the whole of India.

(*3*) It shall come into force on such date[1] as the Central Government may, by notification in the Official Gazette, appoint:

Provided that different dates may be appointed for different provisions of this Act and any reference in any such provision to the commencement of this Act shall be construed as a reference to the coming

Figure: First page of the Biological Diversity Act 2002

The act sets up the following bodies:

-   National Biodiversity Authority

-   State Biodiversity Boards

-   Biodiversity Management Committees

The act holds that biodiversity related activities cannot be undertaken without the approval of the National Biodiversity Authority. Any person or organization must apply to the authority before applying for a patent or intellectual property protection in this area. The authority is to advise the central government on matters of biodiversity, sustainable use of its components and equitable sharing of benefits out of usage of biological resources.

The goals of this act include protection of traditional knowledge, prevention of biopiracy, and prohibition of people and companies from applying for patents without government permission.

The act also has provisions for the development of national and state plans for the conservation of biodiversity, state notification and conservation of biological diversity areas, and notification of endangered species. It also seeks to ensure that benefits from available biological resources, their by-products, knowledge and related practices are equitably shared between the person or organization applying for acquiring such benefits (such as patents) and the local bodies involved.

7.9 Conclusion

India's Wildlife Protection Act of 1972, along with the Forest Conservation Act 1980 as well as Biodiversity Act 2002, have played a pivotal role in the conservation of wildlife by establishing a legal framework for protecting species and biodiversity in India, regulating hunting and trade, and creating protected areas. While challenges remain, the Act's provisions and enforcement mechanisms have significantly contributed to the recovery and protection of many species. Continued efforts to strengthen implementation, address emerging threats, and engage communities will be crucial in advancing wildlife conservation in India and ensuring the long-term survival of its rich biodiversity.

Chapter 8: National Wildlife Action Plan

India's biodiversity is among the richest in the world, but its wildlife faces continuous threats from habitat loss, poaching, human-wildlife conflict, and climate change. To address these challenges, the Government of India formulated the National Wildlife Action Plan (NWAP), an essential policy framework aimed at conserving the country's wildlife. First introduced in 1983, the NWAP has since undergone updates to reflect changing environmental and socio-economic conditions. The latest version, the Third National Wildlife Action Plan (2017–2031), presents a more comprehensive and forward-thinking approach.

8.1 Evolution of the NWAP

The first NWAP (1983-2001) was launched with a strong focus on creating protected areas, such as national parks and wildlife sanctuaries, and addressing immediate threats to wildlife species, particularly those endangered or vulnerable. The second NWAP (2002-2016) expanded upon these ideas, with greater attention on community participation, strengthening wildlife laws, and tackling issues like poaching and illegal wildlife trade.

However, it was clear that conservation efforts needed to be more dynamic, considering new global challenges like climate change, sustainable development, and human-wildlife conflicts. The Third National Wildlife Action Plan (2017–2031) is built on these changing realities, offering a more holistic and inclusive approach to wildlife conservation.

8.2 Key Objectives of the Third NWAP (2017-2031)

The key objectives include the following:

Strengthening and Expanding Protected Areas: Protected areas remain at the core of India's conservation strategy. The Third NWAP advocates for the expansion and better management of these areas. It emphasizes increasing the number of national parks, sanctuaries, and biosphere reserves and creating buffer zones around these habitats to reduce human interference.

Mainstreaming Wildlife Conservation in National Development: Another key feature of the Third NWAP is the recognition that wildlife conservation cannot be seen in isolation from national development. The plan encourages integrating conservation goals with sectors like infrastructure, agriculture, tourism, and industry. It also stresses the importance of sustainable land-use practices and environmental impact assessments in all development projects.

Climate Change Mitigation and Adaptation: With global warming leading to unpredictable changes in ecosystems, the Third NWAP emphasizes the need to assess climate change's impact on wildlife habitats and species. Strategies for ecosystem-based adaptation, such as restoring degraded ecosystems, managing forests, and protecting key wildlife corridors, are included as long-term solutions.

Addressing Human-Wildlife Conflict: Human-wildlife conflict is a growing problem in India, especially in areas where populations are dense and agricultural fields border wildlife habitats. The plan outlines strategies to reduce conflicts by improving community awareness, compensation schemes, and developing better conflict mitigation techniques like early warning systems and wildlife corridors.

Combating Poaching and Illegal Wildlife Trade: Poaching and the illegal wildlife trade continue to pose significant threats to many species in India. The NWAP prioritizes strengthening enforcement mechanisms, building capacity for frontline staff, and engaging local communities to support anti-poaching measures. International cooperation is also a focus, particularly in addressing transboundary wildlife crime.

Community Involvement and Livelihood Generation: The plan stresses the importance of involving local communities in conservation efforts. By promoting sustainable livelihood options like eco-tourism, handicrafts, and agroforestry, it encourages communities to see wildlife conservation as beneficial to their economic well-being. This participatory approach not only strengthens conservation initiatives but also reduces instances of human-wildlife conflict.

Conservation of Marine Ecosystems and Coastal Areas: For the first time, the Third NWAP emphasizes marine conservation, recognizing the critical importance of coastal and marine ecosystems. Protection of coral reefs, mangroves, and endangered marine species like turtles and dugongs is a priority, reflecting India's unique biodiversity both on land and in its waters.

8.3 Implementation and Challenges

The success of the NWAP largely depends on the effective implementation of its policies. The plan calls for the collaboration of various stakeholders, including government bodies, non-governmental organizations (NGOs), research institutions, and local communities. Financial resources, political will, and public awareness are crucial to achieving the desired outcomes.

However, despite the comprehensive vision laid out in the NWAP, challenges remain. Rapid urbanization, population pressure, and conflicting interests between development and conservation often complicate implementation. Moreover, while the plan is progressive in its approach, the execution on the ground requires strong governance and monitoring systems.

8.4 Conclusion

India's National Wildlife Action Plan is a vital tool in addressing the complex challenges faced by its wildlife and habitats. With its forward-looking approach, particularly in terms of integrating development with conservation, the Third NWAP provides a roadmap for sustainable wildlife management.

Chapter 9: Conservation Organizations and Agencies for Wildlife in India

India is home to an astonishing array of wildlife species and habitats, from the dense tropical forests of the Western Ghats to the vast grasslands of the Indo-Gangetic plains. Protecting this rich biodiversity requires coordinated efforts from government agencies, non-governmental organizations (NGOs), research institutions, and universities. This chapter will examine the key players in wildlife conservation in India, exploring the roles and contributions of government bodies, NGOs, and academic institutions

9.1 Government Agencies

Government bodies play a crucial role in developing, implementing, and monitoring wildlife conservation policies across India. The following are the most important government agencies dedicated to wildlife protection and management.

Ministry of Environment, Forest and Climate Change (MoEFCC): The Ministry of Environment, Forest and Climate Change (MoEFCC) is the primary government agency responsible for framing policies related to environmental conservation, forestry, and wildlife protection. Its role encompasses formulating legislation, setting conservation goals, and overseeing the implementation of conservation projects nationwide.

The MoEFCC administers several landmark wildlife protection initiatives, including:

- The Wildlife Protection Act, 1972, which provides legal protection to many threatened species.

- Management of protected areas such as national parks, wildlife sanctuaries, and biosphere reserves.

- Supervision of initiatives like Project Tiger and Project Elephant aimed at conserving specific endangered species.

National Tiger Conservation Authority (NTCA): The National Tiger Conservation Authority (NTCA) was established in 2005 under the Wildlife Protection Act with the specific mandate to strengthen tiger conservation. India is home to nearly 70% of the world's tigers, making their protection a top priority for conservation. The NTCA oversees the management of 50+ tiger reserves across the country, ensuring proper implementation of Project Tiger and coordinating anti-poaching measures, habitat management, and human-tiger conflict mitigation strategies.

Wildlife Crime Control Bureau (WCCB): The Wildlife Crime Control Bureau (WCCB) is a specialized multi-disciplinary agency tasked with fighting wildlife-related crimes, including poaching, trafficking, and illegal trade of wildlife and their products. The WCCB collaborates with state and central law enforcement agencies to tackle the growing menace of organized wildlife crime, and it plays an active role in ensuring India's compliance with international wildlife trade regulations under the **Convention on International Trade in Endangered Species (CITES).**

Central Zoo Authority (CZA): The Central Zoo Authority (CZA) is responsible for the regulation of zoos across India. It sets standards and policies for the management of captive wildlife, with an emphasis on education, conservation breeding, and research. Through these initiatives, the CZA promotes ex-situ conservation, enabling zoos to serve as hubs for endangered species' breeding programs.

9.2 Non-Governmental Organizations (NGOs)

Non-governmental organizations (NGOs) have emerged as critical partners in wildlife conservation. Many of these organizations work closely with local communities, conduct scientific research, and engage in advocacy efforts to shape conservation policy. Some prominent NGOs include the following:

World Wide Fund for Nature (WWF-India): WWF-India, the Indian chapter of the global WWF network, is one of the largest and most influential conservation organizations in the country. Since its establishment in 1969, WWF-India has been involved in a range of activities including species conservation, habitat restoration, environmental education, and combating wildlife trafficking.

WWF-India's key programs include:

- Protecting endangered species such as the tiger, elephant, snow leopard, and rhinoceros.

- Conserving biodiversity-rich landscapes like the Terai Arc and the Sundarbans.

- Promoting sustainable livelihoods and reducing human-wildlife conflict through community engagement.

Wildlife Trust of India (WTI): Founded in 1998, the Wildlife Trust of India (WTI) focuses on innovative and practical solutions to conservation challenges. WTI works in collaboration with government bodies, local communities, and other NGOs to protect endangered species, secure critical habitats, and promote conflict mitigation. One of WTI's most notable projects is the Rapid Action Projects (RAPs), which address immediate and critical wildlife conservation needs in the field, such as rescuing injured or displaced animals and assisting in wildlife law enforcement. Other key initiatives include the Gaj Yatra, aimed at securing corridors for elephant movement, and efforts to protect the Great Indian Bustard and other lesser-known threatened species.

Bombay Natural History Society (BNHS): Established in 1883, the Bombay Natural History Society (BNHS) is one of the oldest conservation NGOs in India. BNHS is a leader in ornithological research and has made significant contributions to avian conservation across India. Over the decades, BNHS has expanded its focus to encompass the broader protection of ecosystems and wildlife. It conducts long-term ecological studies, educates the public on conservation, and advocates for robust environmental policies.

9.3 Research Institutions

Research institutions in India are essential in advancing scientific knowledge and providing a foundation for evidence-based conservation strategies. These academic bodies carry out research on ecology, wildlife biology, conservation genetics, and climate change impacts, among other topics, helping to guide effective conservation policy and practice.

Wildlife Institute of India (WII): The Wildlife Institute of India (WII), based in Dehradun, is a premier institution dedicated to wildlife research, training, and education. Established in 1982, WII plays a pivotal role in shaping conservation strategies through its cutting-edge research on wildlife management, biodiversity conservation, and habitat restoration. The institute provides specialized training to wildlife managers, forest officers, and researchers, ensuring that they are equipped with the necessary skills to address India's complex conservation challenges.

Salim Ali Centre for Ornithology and Natural History (SACON): Named after the eminent ornithologist Dr. Salim Ali, SACON focuses on the study of birds and their habitats. The center conducts research on avian ecology, conservation biology, and environmental monitoring. SACON plays a critical role in promoting bird conservation across India, including efforts to protect migratory species and wetland habitats.

9.4 Indian Universities

Many universities in India, such as Jawaharlal Nehru University (JNU), Indian Institute of Science (IISc), and Aligarh Muslim University (AMU), offer specialized programs in ecology, environmental science, and wildlife conservation. These academic institutions provide research opportunities for students, contribute to policy-making through expert consultations, and often collaborate with government bodies and NGOs on conservation projects.

9.5 Conclusion

The conservation of wildlife in India is a collaborative effort involving government agencies, NGOs, research institutions, and universities. While the government sets the legal framework and implements large-scale programs, NGOs play an active role in advocacy, fieldwork, and community involvement. Research institutions provide the necessary scientific knowledge to guide conservation strategies, while universities help train the next generation of conservationists. Together, these organizations form a robust network that works toward preserving India's unparalleled biodiversity.

Chapter 10: The Role of Indian Courts in Wildlife Conservation

India's legal system has played a crucial role in wildlife conservation, with courts at various levels serving as guardians of environmental law and champions for the protection of biodiversity. Through landmark judgments, judicial interventions, and the enforcement of environmental laws, Indian courts have significantly influenced wildlife conservation efforts in the country. This chapter examines the role of Indian courts in wildlife conservation, highlighting key judicial actions, legal principles, and the impact of court rulings on environmental protection.

10.1 Judicial Oversight and Environmental Protection

Indian courts have been instrumental in shaping and enforcing wildlife conservation policies. The judiciary's role extends beyond interpreting laws to actively participating in the protection of the environment. This proactive approach is grounded in the Indian Constitution, which enshrines the right to a clean and healthy environment as part of the fundamental right to life under Article 21. Courts have leveraged this constitutional mandate to ensure that environmental and wildlife conservation concerns are addressed.

10.2 Fundamental Right to Environment

The Supreme Court of India has established that the right to a clean environment is an integral part of the right to life under Article 21 of the Indian Constitution. This principle was first articulated in the landmark case of Subhash Kumar v. State of Bihar (1991), where the Court held that the right to a clean environment is a fundamental right. This ruling set a precedent for subsequent cases where environmental degradation and wildlife protection were at stake.

10.3 Judicial Activism and Public Interest Litigation

The Indian judiciary, particularly the Supreme Court and High Courts, have embraced judicial activism to address environmental issues. Public Interest Litigation (PIL) has become a powerful tool for citizens and non-governmental organizations (NGOs) to seek judicial intervention on environmental and wildlife conservation matters. PILs allow the courts to address issues that may not be brought before them through traditional litigation, thereby expanding the scope of judicial oversight.

10.4 Landmark Judgments in Wildlife Conservation

Several landmark judgments have demonstrated the Indian judiciary's commitment to wildlife conservation. These cases have addressed a range of issues, from habitat destruction and poaching to the implementation of conservation policies.

T. N. Godavarman Thirumulpad v. Union of India (1997): This case, commonly known as the Forest Case, marked a turning point in forest and wildlife conservation jurisprudence. The Supreme Court issued a series of orders aimed at halting deforestation and regulating forest management practices. The Court's directives included a ban on the clearance of forest land for non-forest purposes and the establishment of a national advisory committee to oversee forest conservation efforts.

Centre for Environment Law v. Union of India (2002): In this case, the Supreme Court addressed the issue of illegal mining in forest areas. The Court ruled that mining activities in forest land could only proceed with the explicit approval of the Ministry of Environment and Forests (MoEF). This judgment emphasized the need for strict adherence to environmental regulations and the protection of wildlife habitats from industrial activities.

The People's Union for Civil Liberties v. Union of India (2004): This case involved the illegal killing of tigers in the Sariska Tiger Reserve. The Supreme Court's intervention led to the closure of the Reserve for tourism and the implementation of measures to restore its tiger population. The Court also ordered the relocation of human settlements from the Reserve's core area, highlighting its role in safeguarding endangered species.

Wildlife Trust of India v. Union of India (2012): This case focused on the conservation of elephants and their migratory routes. The Supreme Court issued a directive for the creation of elephant corridors and the establishment of guidelines to mitigate human-elephant conflict. The Court's ruling emphasized the importance of preserving wildlife corridors for the safe movement of elephants across their traditional habitats.

10.5 Enforcement of Environmental Laws

The enforcement of wildlife protection laws is a critical aspect of the judiciary's role in conservation. Indian courts have been involved in ensuring that laws such as the Wildlife Protection Act (1972) and the Forest Conservation Act (1980) are implemented effectively. Judicial orders often compel government agencies to take action against violators and to ensure compliance with environmental regulations.

10.6 Impact and Challenges

The involvement of Indian courts in wildlife conservation has led to significant progress in protecting India's natural heritage. Judicial rulings have resulted in the preservation of critical habitats, the enforcement of conservation laws, and the promotion of sustainable practices. However, challenges remain, including the need for consistent implementation of court orders, coordination among various stakeholders, and addressing the socio-economic impacts of conservation measures on local communities.

Effectiveness of Court Orders: While judicial orders have had a positive impact, the effectiveness of court rulings often depends on the commitment and capacity of government agencies to implement them. Delays in enforcement, lack of resources, and bureaucratic hurdles can undermine the intended outcomes of judicial interventions.

Integration with Conservation Strategies: Courts often work in tandem with other conservation strategies, including policy reforms, community engagement, and international cooperation. The integration of judicial measures with broader conservation efforts can enhance the overall effectiveness of wildlife protection initiatives.

10.7 Conclusion

Indian courts have played a pivotal role in wildlife conservation by interpreting environmental laws, addressing legal challenges, and ensuring the protection of India's rich biodiversity. Through landmark judgments, judicial activism, and the enforcement of environmental regulations, the judiciary has contributed significantly to the conservation of wildlife and their habitats. As India continues to face new conservation challenges, the role of the judiciary will remain critical in shaping and sustaining efforts to protect the country's natural heritage.

Chapter 11: Biodiversity Hotspots in India

In this chapter we briefly survey some important biodiversity hotspots in India.

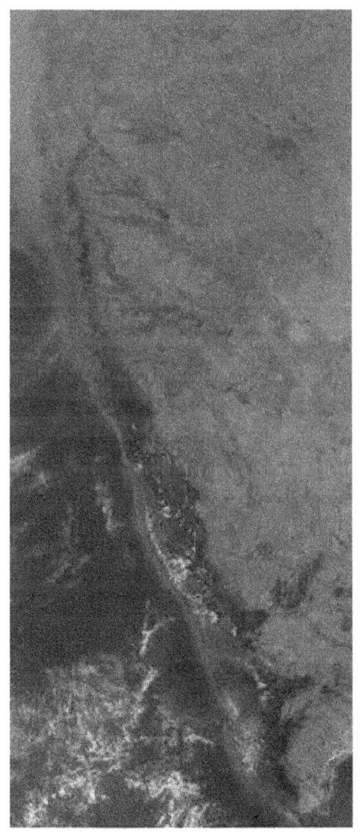

Figure: Satellite photo of the Western Ghats and the Coasts of Malabar and Konkan. By NASA - File:South_India_satellite.jpg, Public Domain, https://commons.wikimedia.org/w/index.php?curid=45555351

Figure: Illustration of biodiversity in Western ghats

11.1 Western Ghats

The Western Ghats, also known as the Sahyadri mountain range, stretch along the western coast of India, spanning six states: Gujarat, Maharashtra, Goa, Karnataka, Kerala, and Tamil Nadu. This region, one of the eight "hottest" biodiversity hotspots in the world, is not just a geographical feature but a cradle of life, harboring a treasure trove of flora and fauna. The mountain range, covering an area of about 160,000 square kilometers, is a UNESCO World Heritage Site, recognized for its unique ecosystems and the variety of life forms it sustains.

Ecological Significance: The Western Ghats are home to some of the most pristine and complex ecosystems on the planet. With tropical rainforests, montane forests, grasslands, and freshwater systems, the region supports an immense variety of biodiversity. Over 7,400 species of flowering plants, 139 mammal species, 508 bird species, and 179 amphibian species thrive in this unique landscape. Many of these species are endemic, meaning they are found nowhere else on Earth.

The diversity of life in the Western Ghats can be attributed to its varied topography and climate. The heavy monsoon rains that lash the region for several months of the year create a rich habitat for both plant and animal species. The Ghats also act as a natural barrier, capturing moisture-laden winds from the Arabian Sea, which feed the region's rivers and support extensive forests. These forests not only serve as a refuge for wildlife but are also vital for the livelihood of millions of people.

Endemic Species and Conservation Challenges: The Western Ghats are a haven for many species that have adapted to the region's specific microclimates. Some well-known endemic species include the Nilgiri tahr, the lion-tailed macaque, and the Malabar pied hornbill. Amphibians, especially frogs, exhibit the highest levels of endemism, with new species still being discovered in remote pockets of the Ghats.

However, the richness of biodiversity also makes the region highly vulnerable to threats. Deforestation, agricultural expansion, and urbanization have significantly impacted the Western Ghats over the past few decades. Hydroelectric projects and mining have further fragmented the landscape, affecting the habitats of many species. One of the most pressing conservation challenges is balancing development with the protection of fragile ecosystems.

Role in Climate Regulation and Ecosystem Services: Apart from biodiversity, the Western Ghats play a crucial role in regulating the climate of peninsular India. The forests here act as carbon sinks, mitigating climate change by absorbing carbon dioxide. They also help in maintaining the hydrological cycle, feeding rivers such as the Godavari, Krishna, and Cauvery, which are vital for agriculture and drinking water in the southern states.

The ecosystem services provided by the Western Ghats are immense. They support agriculture, provide water resources, and offer materials such as timber, medicinal plants, and other forest products. Moreover, the Western Ghats are a cultural and spiritual haven, with several ancient temples, monasteries, and indigenous communities that have lived in harmony with nature for centuries.

Conservation Efforts and the Road Ahead: Recognizing the importance of this region, various initiatives have been undertaken to conserve its biodiversity. Protected areas, national parks, and wildlife sanctuaries have been established, including the Silent Valley National Park, Periyar Wildlife Sanctuary, and Kalakkad Mundanthurai Tiger Reserve. International attention through its designation as a UNESCO World Heritage Site has also raised awareness of its conservation needs.

Yet, the road ahead remains challenging. Involving local communities in conservation efforts, promoting sustainable agricultural practices, and preventing illegal logging and mining are crucial for ensuring that the Western Ghats continue to thrive. Conservation efforts must be holistic, balancing economic needs with ecological integrity, and ensuring that future generations inherit a landscape that remains rich in biodiversity.

The Western Ghats are a testament to India's natural heritage. Protecting them is not only vital for the flora and fauna that call them home but for the millions of people who depend on the ecological services they provide.

Figure: Illustration of the biodiversity of Eastern Himalayas

11.2 The Eastern Himalayas

The Eastern Himalayas, straddling the northeastern region of India and parts of Bhutan, Nepal, and Tibet, are among the most important biodiversity hotspots in the world. This region, which encompasses states like Sikkim, Arunachal Pradesh, and parts of Assam, is renowned for its unique ecosystems and species diversity. Often referred to as the "roof of the world," the Eastern Himalayas are not just an awe-inspiring landscape but a critical zone for conservation efforts.

Ecological Significance: The Eastern Himalayas are a complex mosaic of habitats ranging from tropical forests at the foothills to alpine meadows and glaciers at the higher altitudes. This varied geography creates the perfect conditions for an extraordinary range of life forms. The region is home to over 10,000 plant species, 980 bird species, 300 mammal species, and an astonishing number of amphibians, reptiles, and insects. What makes the Eastern Himalayas particularly special is its high level of endemism, with a large number of species found nowhere else on Earth.

The region's steep altitude gradients play a critical role in this biodiversity. As the Himalayas rise from low-lying plains to towering peaks, they create different climatic zones, each with its distinct species adapted to specific conditions. For example, the temperate forests of Sikkim and Arunachal Pradesh shelter endangered species such as the red panda, while the subalpine and alpine zones host species like the snow leopard and the Himalayan blue sheep.

Endemic and Endangered Species: One of the hallmarks of the Eastern Himalayas is its high concentration of endemic species. Iconic animals like the red panda, Himalayan musk deer, and the clouded leopard are native to this region. Bird enthusiasts flock to the Eastern Himalayas for species like the satyr tragopan, Himalayan monal, and the Bengal florican, all of which are rare and found primarily in this part of the world.

The region is also a botanical paradise, with over 4,000 species of flowering plants, many of which are endemic. Orchids, rhododendrons, and primulas are abundant, especially in Sikkim and Arunachal Pradesh. The Eastern Himalayas are a global center for orchid diversity, with over 750 species recorded here.

However, despite its richness, the Eastern Himalayas are under constant threat. Habitat destruction due to deforestation, shifting agriculture, and infrastructure development poses severe challenges to the conservation of both flora and fauna. Climate change, too, threatens the delicate balance of these ecosystems, altering rainfall patterns and snowmelt, which in turn affect the biodiversity that has evolved over millennia to adapt to specific climatic conditions.

Cultural and Ecological Interactions: The Eastern Himalayas are not only rich in biodiversity but are also a cultural and spiritual hub. The indigenous communities living here, such as the Lepchas, Sherpas, and Monpas, have coexisted with nature for centuries, relying on traditional knowledge for sustainable living. Their reverence for the environment is reflected in their customs, beliefs, and practices, which have helped preserve biodiversity in the region.

Sacred forests, lakes, and mountains are protected by these communities, acting as natural reserves. In Sikkim, for instance, the state government has declared the entire region an organic state, promoting agriculture that is in harmony with nature. Such initiatives showcase how human activities, when aligned with conservation goals, can help protect biodiversity rather than deplete it.

Conservation Challenges and Opportunities: Despite its ecological and cultural significance, the Eastern Himalayas face several challenges. Deforestation driven by agricultural expansion, logging, and urbanization is one of the most pressing issues. The region's fragile ecosystems are highly sensitive to these disturbances, which can lead to the loss of habitat and species.

Moreover, climate change poses an existential threat to the Eastern Himalayas. The melting of glaciers, erratic monsoon patterns, and increasing temperatures are altering habitats at an alarming rate. Many species that are adapted to specific climatic conditions may find it difficult to survive as temperatures rise and habitats shift upwards, potentially leading to extinction.

In response to these threats, several conservation initiatives have been launched. The creation of protected areas such as the Khangchendzonga National Park in Sikkim and the Namdapha National Park in Arunachal Pradesh has been crucial in safeguarding key ecosystems. International recognition, such as the designation of Khangchendzonga as a UNESCO World Heritage Site, has also brought global attention to the need for conservation in the Eastern Himalayas.

However, conservation efforts must go beyond the creation of protected areas. Integrating local communities into conservation strategies is key to long-term success. By promoting sustainable livelihoods, eco-tourism, and community-led conservation initiatives, the region can balance human development with the preservation of its unique ecosystems.

The Road Ahead: The Eastern Himalayas are not just a biodiversity hotspot; they are a global ecological treasure. Their conservation is not only important for the species that live here but also for the millions of people who depend on the ecosystem services they provide, including water regulation, carbon sequestration, and agriculture.

Efforts to preserve the Eastern Himalayas must be multifaceted, incorporating both local and international stakeholders. With the right blend of policy, science, and community engagement, this fragile and beautiful region can continue to thrive, ensuring that future generations can inherit the wealth of life it shelters.

Figure: Illustration of the biodiversity in Sunderbans

11.3 The Sunderbans

The Sundarbans, located in the coastal regions of West Bengal, India, and extending into Bangladesh, form the largest contiguous mangrove forest in the world. A UNESCO World Heritage Site, this remarkable region is also recognized as a critical biodiversity hotspot due to its unique ecosystem, which supports a rich diversity of flora and fauna. The Sundarbans are not only crucial for biodiversity but also serve as a vital buffer against natural disasters like cyclones, while being home to millions of people who rely on its resources.

Ecological Significance: The Sundarbans derive their name from the "Sundari" tree, one of the dominant mangrove species in the area. These mangroves are the lifeline of the ecosystem, supporting a complex network of tidal waterways, mudflats, and small islands. The forest plays a critical role in stabilizing the coastal ecosystem and preventing soil erosion by acting as a natural barrier between the Bay of Bengal and the mainland.

What sets the Sundarbans apart from other biodiversity hotspots is its status as a tidal halophytic (salt-tolerant) mangrove ecosystem. This unique environment nurtures an array of species that have adapted to its saline conditions. The Sundarbans are renowned for housing the largest population of the Bengal tiger, particularly the elusive Royal Bengal tiger, which has evolved to swim between islands in search of prey. These tigers, known as "swimming tigers," are emblematic of the region's wild and untamed nature.

Endemic and Endangered Species: The Sundarbans are a haven for wildlife, hosting over 260 bird species, 42 mammal species, 35 reptile species, and numerous species of fish, amphibians, and invertebrates. Some iconic species that thrive here include the saltwater crocodile, Indian python, river terrapin, and the Gangetic dolphin. The region is also home to several endangered species like the masked finfoot and the Irrawaddy dolphin, as well as migratory birds that arrive here in the winter months.

The mangroves of the Sundarbans support an intricate food web, starting with the detritus from fallen mangrove leaves, which sustains crabs, fish, and mollusks. These, in turn, are preyed upon by larger species such as otters, birds of prey, and, at the top of the food chain, the Bengal tiger.

However, the Sundarbans' ecosystem is fragile and faces numerous threats. Rising sea levels due to climate change, coupled with increased salinity and habitat loss, are endangering many species. Illegal poaching, overfishing, and human-wildlife conflict—particularly involving tigers—also contribute to the pressures faced by this unique habitat.

Ecosystem Services and Livelihoods: The Sundarbans are not only vital for wildlife but also for the millions of people who live in and around the area. The mangroves provide critical ecosystem services such as carbon sequestration, protection from cyclones and storm surges, and support for fisheries, which are an essential source of food and income for local communities. The mangrove roots act as breeding grounds for fish and shrimp, sustaining both small-scale and commercial fisheries.

The forest is also a source of honey, wood, and other forest products. Local communities have traditionally relied on the Sundarbans for their livelihoods, following a sustainable approach that includes honey collection, crab fishing, and small-scale agriculture. These practices, while beneficial, must be managed carefully to avoid overexploitation of the region's delicate resources.

Conservation Challenges: Despite its ecological and economic importance, the Sundarbans face immense conservation challenges. Climate change is arguably the biggest threat, as rising sea levels and increasing salinity are gradually submerging the low-lying islands. This not only threatens the habitat of the wildlife but also endangers the lives and livelihoods of the local population.

Another significant challenge is human-wildlife conflict. The Bengal tiger, while iconic, often comes into conflict with people living near the forest. Villagers who venture into the forest for fishing or collecting honey are sometimes attacked by tigers, leading to retaliatory killings. Managing this conflict through community engagement and alternative livelihood programs is crucial for the long-term survival of both tigers and people.

Additionally, the expansion of aquaculture, particularly shrimp farming, has led to the degradation of mangrove forests, further disrupting the ecosystem. Unsustainable fishing practices and pollution from nearby urban areas also contribute to the pressures on the Sundarbans.

Conservation Efforts and the Future: Several conservation efforts have been implemented to protect the Sundarbans, both at the national and international levels. The Indian government, in collaboration with organizations like the World Wide Fund for Nature (WWF) and the International Union for Conservation of Nature (IUCN), has designated large parts of the Sundarbans as protected areas, including the Sundarbans National Park and Tiger Reserve.

Community-based conservation initiatives are also gaining momentum, with efforts to involve local populations in sustainable resource management and eco-tourism. These initiatives aim to provide alternative livelihoods that reduce dependence on the forest, such as promoting honey cultivation, fish farming, and craft production, while also educating people on the importance of conserving the fragile ecosystem.

The future of the Sundarbans depends on addressing climate change, habitat conservation, and the needs of the local communities. With rising sea levels threatening to submerge parts of the mangrove forest, urgent global and local action is needed to mitigate the effects of climate change, while improving conservation efforts through habitat restoration and better management of resources.

The Road Ahead: The Sundarbans represent a rare convergence of biodiversity and human culture, where survival is deeply interconnected. As one of India's most significant biodiversity hotspots, its protection is essential not just for the iconic Bengal tiger but for the countless species and people who depend on it. Conservation efforts must be strengthened, with a focus on resilience against climate change, community involvement, and the protection of natural resources. Only through a holistic approach can this extraordinary landscape continue to thrive for generations to come.

11.4 Conclusion

In this chapter we have discussed some important conservation hotspots of India.

Chapter 12: Conservation of Some Flagship Species of India

India is home to some of the most iconic and majestic species on Earth, many of which are classified as "flagship species." These animals serve as symbols of biodiversity conservation, capturing the public's imagination and galvanizing efforts to protect their habitats and the ecosystems in which they live. Four of the most prominent flagship species in India are the Bengal tiger, Asiatic lion, Indian elephant, and the one-horned rhinoceros. Each of these species plays a critical role in maintaining ecological balance and represents the rich natural heritage of the country.

Figure: An adult male Royal Bengal tiger in Kanha National Park, Madhya Pradesh, India. By Seemaleena - Own work, CC BY-SA 4.0, https://commons.wikimedia.org/w/index.php?curid=49329017

12.1 Bengal Tiger (*Panthera tigris tigris*)

The Bengal tiger, often referred to as the "National Animal of India," is one of the most powerful symbols of wildlife conservation in the world. Found primarily in the forests of India, Bangladesh, Bhutan, and Nepal, the Bengal tiger is the largest tiger subspecies and one of the most charismatic megafauna in the animal kingdom. India is home to the largest population of Bengal tigers, with an estimated 3,000 individuals living in the wild.

Habitat and Ecology: The Bengal tiger thrives in a variety of habitats, including tropical rainforests, mangrove swamps (notably in the Sundarbans), grasslands, and dry deciduous forests. As an apex predator, the Bengal tiger plays a crucial role in controlling prey populations, maintaining the balance of ecosystems. Its primary prey consists of deer species like the sambar and chital, as well as wild boar and occasionally, smaller mammals.

Conservation Status and Challenges: Listed as Endangered on the IUCN Red List, the Bengal tiger has faced significant threats over the past century due to poaching, habitat loss, and human-wildlife conflict. The demand for tiger skins and body parts for traditional medicine has led to rampant poaching. However, India's "Project Tiger," launched in 1973, has been instrumental in increasing tiger numbers through the creation of tiger reserves and better protection of habitats. Reserves like Bandhavgarh, Kanha, and Ranthambore have become global success stories in tiger conservation.

Figure: Indian lion male (Panthera leo persica) at Gir National Park. By Shanthanu Bhardwaj, CC BY-SA 2.0, https://commons.wikimedia.org/w/index.php?curid=26124541

12.2 Asiatic Lion (*Panthera leo persica*)

The Asiatic lion, distinct from its African counterpart, once roamed across the Middle East and parts of India but is now restricted to a single population in the Gir Forest of Gujarat. These lions are slightly smaller than African lions, with males having shorter manes. Their survival story, from near extinction to steady recovery, is a testament to India's focused conservation efforts.

Habitat and Ecology: Unlike African lions that live in open savannahs, Asiatic lions inhabit dry deciduous forests and grasslands in Gir. The Gir Forest National Park is the last refuge of these lions, supporting a population of around 700 individuals. Asiatic lions live in prides but tend to be more territorial and have smaller pride sizes than their African counterparts. They primarily prey on herbivores such as chital, nilgai, and wild boar, but they occasionally target livestock, leading to conflict with local communities.

Conservation Status and Challenges: By the early 20th century, Asiatic lions were on the brink of extinction due to hunting and habitat destruction. Today, they are listed as Endangered, but their population has steadily increased due to the efforts of the Gujarat Forest Department and local communities. However, the lack of genetic diversity in the small, isolated population poses a significant challenge, as does the risk of disease outbreaks and habitat fragmentation. Plans to relocate some lions to Kuno National Park in Madhya Pradesh to establish a second population have faced delays, but remain vital for the species' long-term survival.

Figure: Indian elephant bull in Bandipur National Park. By Yathin S Krishnappa - Own work, CC BY-SA 3.0, https://commons.wikimedia.org/w/index.php?curid=24916395

12.3 Indian Elephant (*Elephas maximus indicus*)

The Indian elephant, a subspecies of the Asian elephant, is one of the most revered animals in India, playing a significant role in Indian culture, religion, and history. Elephants are considered symbols of wisdom and strength, often associated with Lord Ganesha, the Hindu god of wisdom and prosperity. These gentle giants, however, are under threat and serve as flagship species for habitat conservation.

Habitat and Ecology: Indian elephants inhabit diverse ecosystems ranging from tropical and subtropical forests to grasslands and scrublands across India, especially in states like Karnataka, Kerala, Assam, and West Bengal. They are highly social animals, living in matriarchal herds and covering vast distances in search of food and water. Indian elephants are ecosystem engineers, shaping the landscape by trampling vegetation, dispersing seeds, and creating waterholes, which in turn support other wildlife.

Conservation Status and Challenges: Listed as Endangered, the Indian elephant faces significant challenges due to habitat loss, fragmentation, and human-elephant conflict. As agriculture and human settlements encroach upon elephant corridors, these majestic creatures often come into conflict with humans, leading to crop damage, property destruction, and sometimes loss of life. Conservation initiatives such as "Project Elephant," launched in 1992, aim to mitigate these conflicts by protecting elephant corridors, establishing elephant reserves, and encouraging local community involvement in conservation efforts. Reserves like the Nilgiri Biosphere and Periyar Tiger Reserve are key habitats for India's elephants.

Figure: An Indian rhinoceros in Kaziranga National Park. By Mayank1704 - Own work, CC BY-SA 4.0, https://commons.wikimedia.org/w/index.php?curid=49547280

12.4 One-Horned Rhinoceros (*Rhinoceros unicornis*)

The one-horned rhinoceros, also known as the Indian rhinoceros, is one of the most iconic and endangered animals in South Asia. Known for its thick, armor-like skin and single horn, this species once roamed across the entire northern Indian subcontinent. Today, it is primarily found in northeastern India and parts of Nepal, with Assam's Kaziranga National Park serving as the stronghold for the species.

Habitat and Ecology: The one-horned rhinoceros inhabits floodplains, grasslands, and swamps in the foothills of the Himalayas. These solitary herbivores are well-adapted to grazing on tall grasses, shrubs, and aquatic plants. Their large size and voracious appetite help shape the grassland ecosystems, maintaining open spaces that benefit other herbivores.

Conservation Status and Challenges: Once hunted to near extinction for their horns, which are highly valued in traditional medicine, the one-horned rhinoceros has made a remarkable recovery due to stringent conservation measures. In the early 20th century, fewer than 200 rhinos remained. Today, thanks to efforts like anti-poaching patrols and habitat protection in reserves like Kaziranga, Manas, and Pobitora, the population has rebounded to over 3,700 individuals. However, the species remains Vulnerable, with poaching and habitat degradation posing ongoing threats. Additionally, the floodplains they inhabit are highly vulnerable to seasonal flooding, which can impact both rhinos and their habitat.

12.5 Conclusion

The Bengal tiger, Asiatic lion, Indian elephant, and one-horned rhinoceros are not just wildlife species; they are emblematic of India's natural heritage and serve as flagships for broader conservation efforts. Each of these animals plays a critical role in maintaining the ecological balance of their habitats. However, they also face significant threats from poaching, habitat destruction, and human-wildlife conflict. Conservation initiatives, including national programs like Project Tiger and Project Elephant, have made considerable progress, but ongoing efforts are required to secure a future for these majestic creatures.

Chapter 13: Conservation of India's River and Sea Fish and Ecology

India is home to a vast diversity of aquatic ecosystems, including rivers, lakes, estuaries, and a coastline that stretches over 7,500 kilometers along the Arabian Sea and the Bay of Bengal. These aquatic habitats support a rich diversity of fish species and marine life, many of which are integral to the livelihoods of millions of people. However, like India's terrestrial ecosystems, these aquatic environments face numerous threats due to human activities. This chapter focuses on the importance of conserving India's river and sea fish, as well as the broader conservation of marine and river ecosystems.

13.1 The Importance of River and Marine Fish in India

India's rivers, lakes, and seas are home to more than 2,000 species of fish, many of which are critical to the country's biodiversity and economy. Fish are an essential source of food and income for millions of people, particularly in coastal and riverine communities. Inland fisheries, especially in the Ganges, Brahmaputra, and Godavari river systems, support traditional fishing livelihoods, while marine fisheries contribute significantly to India's export economy.

Hilsa (Ilish) fish. By Roddur 123 - Own work, CC BY 4.0, https://commons.wikimedia.org/w/index.php?curid=106743571

13.2 Key River and Sea Fish Species in India

Rivers: The hilsa, found in the Ganges and Brahmaputra, is one of India's most prized fish. Other significant freshwater species include the mahseer, rohu, and catfish.

Seas: In the marine ecosystem, species such as Indian mackerel, tuna, pomfret, and sharks are economically and ecologically important.

Beyond their economic importance, fish play critical ecological roles. They help maintain the balance of aquatic ecosystems by controlling algae and insect populations, cycling nutrients, and serving as food for other species.

13.3 Challenges Facing River and Marine Ecology in India

India's river and marine ecosystems face a multitude of threats, many of which are driven by human activity and unsustainable development. These challenges are having profound effects on fish populations and the health of aquatic ecosystems.

Overfishing: Overfishing is a significant threat to both freshwater and marine ecosystems. The increasing demand for fish, driven by population growth and the global seafood market, has led to the depletion of many fish stocks. In rivers, overfishing of species like the hilsa has caused dramatic population declines. In marine waters, fish species such as tuna, sharks, and Indian mackerel are being overexploited.

Illegal, unreported, and unregulated (IUU) fishing further exacerbates this problem, with many fishing practices not adhering to sustainable limits. Destructive fishing methods, such as the use of fine-mesh nets, result in bycatch (the capture of non-target species) and harm juvenile fish populations, reducing their ability to replenish.

Habitat Degradation: Rivers and marine habitats in India are undergoing rapid degradation due to pollution, infrastructure development, and deforestation. In rivers, the construction of dams and hydropower projects, such as those on the Ganges and Brahmaputra, disrupt the natural flow of water, alter sediment patterns, and block the migration routes of fish like the mahseer and hilsa, which rely on free-flowing rivers to breed. Marine ecosystems are also under pressure from coastal development, which destroys critical habitats like mangroves, coral reefs, and seagrass beds that provide breeding and feeding grounds for fish. Mangroves, in particular, serve as nurseries for many commercially valuable fish species, and their loss leads to declining fish populations.

Pollution: Industrial waste, agricultural runoff, and untreated sewage are major sources of pollution in India's rivers and seas. Chemicals like pesticides, heavy metals, and plastics accumulate in water bodies, affecting the health of aquatic life. In rivers, pollution from factories and urban areas leads to oxygen depletion (eutrophication) and harmful algal blooms, making it difficult for fish to survive. In marine ecosystems, plastic pollution is a growing threat, with large amounts of plastic waste entering the oceans and being ingested by fish, turtles, and marine mammals. Oil spills, particularly in busy coastal regions, pose a further risk to marine life.

Climate Change: Climate change is impacting India's aquatic ecosystems in significant ways. Rising temperatures affect the reproductive cycles and migration patterns of fish. In marine environments, ocean acidification (caused by the absorption of excess carbon dioxide) and warming seas are damaging coral reefs and disrupting marine food chains. Sea-level rise, driven by climate change, is threatening coastal habitats such as mangroves and estuaries, which are crucial for fish breeding. In rivers, changes in rainfall patterns and glacial melt are altering water flow, leading to droughts or floods that affect fish populations and their habitats.

13.4 Conservation Efforts for River and Marine Ecosystems

Conserving India's river and marine ecosystems requires a multifaceted approach that includes sustainable management, habitat protection, pollution control, and community engagement.

Sustainable Fisheries Management: Effective fisheries management is essential to prevent overfishing and ensure the long-term sustainability of fish populations. The **Marine Fisheries Regulation Act (MFRA)**, implemented by India's coastal states, regulates fishing activities by enforcing measures such as seasonal fishing bans, size limits for fish catches, and restrictions on the use of certain fishing gear. These regulations aim to protect fish during breeding seasons and allow fish populations to recover.

The promotion of sustainable fishing practices, such as using selective gear to reduce bycatch, implementing no-fishing zones, and promoting aquaculture, can help balance the needs of fishing communities with conservation goals.

Protection of Critical Habitats: Protecting and restoring critical habitats, such as mangroves, coral reefs, and river floodplains, is vital for the conservation of fish and other aquatic species. In recent years, India has implemented several initiatives to restore degraded ecosystems:

- Mangrove Restoration: Programs like the **Mangrove for the Future (MFF)** initiative are working to restore mangroves along India's coasts, particularly in states like West Bengal and Andhra

Pradesh. These efforts help protect coastal communities from storms while providing essential breeding grounds for fish.

- Coral Reef Conservation: Coral reefs, particularly in regions like the Lakshadweep and Andaman and Nicobar Islands, are being protected through marine protected areas (MPAs) and reef restoration projects.

In river ecosystems, the **National Mission for Clean Ganga (NMCG)** aims to reduce pollution, restore natural river flows, and protect aquatic biodiversity in the Ganges and its tributaries. Fish ladders and other structures are being implemented in dammed rivers to allow migratory fish species to bypass barriers and reach their spawning grounds.

13.5 Community Involvement

Local communities, particularly those that depend on fishing for their livelihoods, play a crucial role in conservation. Community-based fisheries management programs empower local fishers to take responsibility for the sustainable use of resources. For example, in Kerala, community initiatives to manage fish stocks in inland waters have led to more sustainable fishing practices and better protection of species like the rohu and catla.

In the Sundarbans, home to the iconic Bengal tiger and critical fisheries, local communities are involved in managing mangrove forests and protecting endangered species. This ensures that conservation efforts are aligned with the needs of local people, creating incentives for long-term environmental stewardship.

13.6 Marine Protected Areas (MPAs)

India has established several marine protected areas (MPAs) to safeguard its marine biodiversity. These protected areas restrict fishing and other extractive activities, providing safe havens for marine species to breed and thrive. MPAs such as the Gulf of Mannar and Malvan Marine

Sanctuary are critical to conserving marine biodiversity and ensuring the recovery of fish stocks.

13.7 Conclusion

India's rivers and seas are vital to its ecological health, economy, and cultural heritage. While these ecosystems face numerous challenges, concerted conservation efforts, involving government agencies, local communities, and international organizations, are making a difference. By prioritizing sustainable fisheries management, habitat restoration, pollution control, and climate adaptation, India can ensure the long-term survival of its rich aquatic biodiversity.

Chapter 14: Conservation of India's Livestock

India's livestock sector plays a pivotal role in its agricultural economy and cultural heritage. As the world's largest producer of milk and a significant producer of meat, eggs, and wool, the country depends heavily on its diverse array of livestock, which includes cattle, buffalo, goats, sheep, poultry, and camels. However, with increasing demands for livestock products and changing agricultural practices, the conservation and preservation of India's living animal resources have become vital to ensuring food security, rural livelihoods, and biodiversity. This chapter explores the importance of conserving livestock diversity, current threats, and the steps being taken to preserve these valuable resources.

14.1 Importance of Livestock in India

Livestock are an integral part of rural life in India. Beyond their role in agriculture, they contribute to income, nutrition, cultural traditions, and ecological sustainability.

Economic Value: India's livestock sector contributes nearly 4% of the country's GDP and provides employment to millions of rural households. Livestock products such as milk, meat, wool, and eggs are critical to the economy. India is the world's largest producer of milk, with breeds like the Murrah buffalo and Gir cattle being renowned for their high milk yields. Poultry farming, particularly with indigenous breeds like the Kadaknath chicken, is also vital to rural economies.

Cultural Significance: Many livestock species hold deep cultural and religious significance in India. Cows are considered sacred in Hinduism, and many communities practice traditional rituals involving livestock. Indigenous livestock breeds, such as the Tharparkar cattle or Marwari horses, have been reared for centuries, forming an essential part of India's cultural identity.

Ecological Role: Indigenous livestock play an important role in maintaining ecological balance. Grazing animals like sheep and goats help manage vegetation, while animals like buffaloes and bullocks provide natural manure, reducing the need for chemical fertilizers. Livestock also contribute to sustainable farming through mixed farming systems, where crops and animals are integrated to enhance productivity and environmental health.

14.2 Threats to India's Livestock Diversity

While India boasts a wealth of livestock biodiversity, several factors pose threats to the conservation of indigenous breeds and traditional animal husbandry practices.

Breed Loss and Genetic Erosion: India is home to a variety of indigenous livestock breeds, each adapted to local environments and conditions. However, the introduction of high-yielding exotic breeds has led to the decline of many native breeds. For instance, the promotion of Holstein Friesian and Jersey cattle for milk production has reduced the populations of indigenous cattle breeds such as the Kankrej and Sahiwal. Genetic erosion occurs when the genetic diversity within livestock populations is reduced, making them less resilient to diseases, climate change, and other environmental stresses.

Crossbreeding and Hybridization: While crossbreeding programs have been successful in increasing milk production, they have also led to the dilution of native breeds. Crossbred animals often lack the hardiness and disease resistance of indigenous breeds, making them more dependent on intensive care and feed inputs. For example, crossbreeding between indigenous and exotic poultry has led to the marginalization of hardy native chicken breeds like the Aseel, which are more resistant to local diseases and better suited to free-range farming.

Changing Agricultural Practices: As agriculture modernizes, the role of traditional livestock practices is diminishing. Mechanization has reduced the need for draft animals like bullocks, and intensive farming practices are replacing mixed farming systems. This shift not only threatens traditional knowledge but also contributes to the loss of livestock breeds that are integral to sustainable farming.

14.3. Conservation Efforts for Livestock in India

Recognizing the importance of preserving livestock biodiversity, the Indian government, research institutions, and non-governmental organizations are taking steps to conserve indigenous breeds and promote sustainable livestock management.

Breed Conservation Programs: The **National Bureau of Animal Genetic Resources (NBAGR)** plays a crucial role in identifying, documenting, and conserving India's livestock breeds. It maintains a comprehensive registry of breeds and works to preserve genetic material through in situ (on-farm) and ex situ (off-farm) conservation methods. The NBAGR has recognized more than 190 indigenous livestock breeds, including cattle, buffalo, goats, sheep, and poultry.

In situ conservation efforts focus on supporting farmers who rear indigenous breeds by providing financial incentives and technical assistance. For example, the **National Livestock Mission (NLM)** promotes the conservation of indigenous breeds by offering subsidies for maintaining purebred herds.

Indigenous Breed Promotion: Programs such as the **Rashtriya Gokul Mission (RGM)** focus specifically on conserving and promoting indigenous cattle breeds. The mission aims to improve the productivity of native breeds through selective breeding and by setting up Gokul Grams (cattle breeding centers) dedicated to indigenous cattle. The program encourages farmers to rear native breeds like Sahiwal, Red Sindhi, and Gir, which are well-adapted to local conditions and are essential for organic farming practices.

Livestock Development and Health Initiatives: Improving animal health is critical to conserving livestock. India has launched several initiatives to provide veterinary care, vaccinations, and disease control programs for livestock. The **National Animal Disease Control Programme (NADCP)** aims to eliminate diseases like foot-and-mouth disease (FMD) and brucellosis, which are major threats to livestock populations. Additionally, the development of feed and fodder resources through initiatives like the National Livestock Mission helps address the issue of malnutrition in livestock, ensuring that indigenous breeds remain healthy and productive.

Sustainable Grazing and Pastoralism: Conserving pastoralist traditions is vital for the preservation of livestock diversity. Pastoralist communities, such as the Raikas of Rajasthan, have traditionally managed livestock like camels and goats through sustainable grazing practices. Supporting these communities through policies that ensure access to grazing lands, water resources, and markets can help preserve both the livestock and the traditional knowledge that has sustained them for generations.

Several NGOs and grassroots organizations are working to revive traditional grazing practices and promote the conservation of breeds like the Banni buffalo and Marwari camel. These efforts not only conserve genetic diversity but also strengthen the livelihoods of pastoralist communities.

14.4 The Future of Livestock Conservation in India

The future of livestock conservation in India hinges on finding a balance between modern agricultural demands and the preservation of traditional, sustainable practices. Key areas of focus for future conservation efforts include:

Promoting Indigenous Breeds for Climate Resilience: As climate change increasingly impacts agriculture, indigenous livestock breeds, which are naturally resilient to local conditions, will become more valuable. Conservation programs must prioritize breeds that can withstand drought, extreme heat, and other climate-related challenges. For example, the Kankrej cattle of Gujarat and Nellore sheep of Andhra Pradesh are known for their ability to thrive in arid environments.

Expanding Research and Genetic Resource Banks: Investing in research to improve the productivity of indigenous breeds through selective breeding and genetic improvement is crucial. Expanding genetic resource banks, which store semen, embryos, and DNA from indigenous breeds, will help preserve genetic diversity and provide options for future breeding programs.

Supporting Sustainable Livestock Farming: Promoting sustainable livestock farming practices, such as organic farming and agroforestry, will help conserve both livestock and the ecosystems they depend on. Programs that integrate livestock into agroecological systems, where animals contribute to soil health and nutrient cycling, can improve food security while preserving biodiversity.

Raising Awareness and Incentivizing Conservation: Public awareness campaigns about the importance of conserving indigenous livestock and traditional animal husbandry practices are essential for generating support among farmers and policymakers. Incentives such as subsidies, financial support for breed conservation, and market access for organic livestock products can motivate farmers to continue rearing indigenous breeds.

14.5 Conclusion

The conservation and preservation of India's livestock are essential not only for maintaining biodiversity but also for supporting rural livelihoods, ensuring food security, and promoting ecological sustainability. By prioritizing the conservation of indigenous breeds, promoting sustainable farming practices, and integrating traditional knowledge with modern science, India can safeguard its living animal resources for future generations.

Chapter 15: Protected Areas and Sanctuaries in India

India is renowned for its rich biodiversity, hosting a wide array of ecosystems that support a diverse range of plant and animal species. To protect this natural heritage, the country has established a network of protected areas, including national parks, wildlife sanctuaries, and biosphere reserves. These protected areas play a crucial role in conserving India's flora and fauna, preserving habitats, and mitigating human-wildlife conflict. This chapter provides an overview of the different types of protected areas, their effectiveness in conservation, and highlights some of the most important protected regions across the country.

India's network of protected areas is categorized into three main types: national parks, wildlife sanctuaries, and biosphere reserves. Each type serves a different purpose in conservation but collectively, they form the backbone of wildlife protection in the country.

15.1 National Parks

National parks are areas designated for the protection of ecosystems, where activities such as agriculture, human settlement, and industrial development are strictly prohibited. They offer the highest level of legal protection to wildlife and habitats. National parks focus on preserving the ecological integrity of areas, often protecting endangered species and unique landscapes.

India's first national park, Jim Corbett National Park, was established in 1936 and remains a key site for tiger conservation. Today, India has over 100 national parks, covering a wide range of habitats including forests, grasslands, and wetlands. National parks like Kaziranga in Assam, Ranthambore in Rajasthan, and Sundarbans in West Bengal are globally recognized for their role in preserving iconic species like the Bengal tiger, one-horned rhinoceros, and saltwater crocodile.

Figure: Wildlife sanctuaries in India. wiki communities, CC BY-SA 4.0 <https://creativecommons.org/licenses/by-sa/4.0>, via Wikimedia Commons

15.2 Wildlife Sanctuaries

Wildlife sanctuaries are areas where wildlife is protected, but human activities, such as controlled grazing, tourism, and resource extraction, are permitted under regulated conditions. These sanctuaries often serve as buffer zones around national parks or as standalone reserves that provide crucial habitats for various species. India has over 500 wildlife sanctuaries, each contributing to the protection of diverse ecosystems.

Sanctuaries like Periyar Wildlife Sanctuary in Kerala and Manas Wildlife Sanctuary in Assam are vital for the conservation of species like the

Indian elephant, tigers, and leopards, while also supporting community livelihoods through eco-tourism and sustainable resource use.

15.3 Biosphere Reserves

Biosphere reserves are large, ecologically significant areas recognized by UNESCO's Man and the Biosphere (MAB) Programme. These reserves integrate biodiversity conservation with sustainable development practices, often involving local communities in managing natural resources. A biosphere reserve consists of three zones: the core zone (strictly protected), the buffer zone (where limited human activity is allowed), and the transition zone (where sustainable development activities are encouraged).

India has 18 designated biosphere reserves, including the Nilgiri Biosphere Reserve (covering parts of Tamil Nadu, Kerala, and Karnataka), the Sundarbans Biosphere Reserve, and the Great Nicobar Biosphere Reserve. These areas are critical not only for the protection of biodiversity but also for scientific research and fostering sustainable practices among local populations.

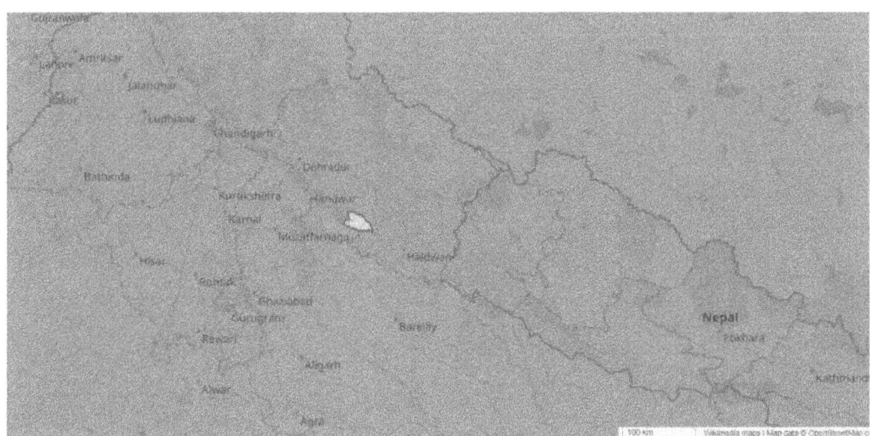

Figure: Location of Corbett National Park in Uttarakhand. Wikimedia Maps.

15.4 Jim Corbett National Park, Uttarakhand

As India's first national park and a flagship site for tiger conservation, Jim Corbett National Park is a symbol of wildlife protection in the country. Located in the foothills of the Himalayas, this park is known for its varied landscapes, including forests, grasslands, and rivers, making it a key habitat for Bengal tigers, leopards, elephants, and over 600 species of birds.

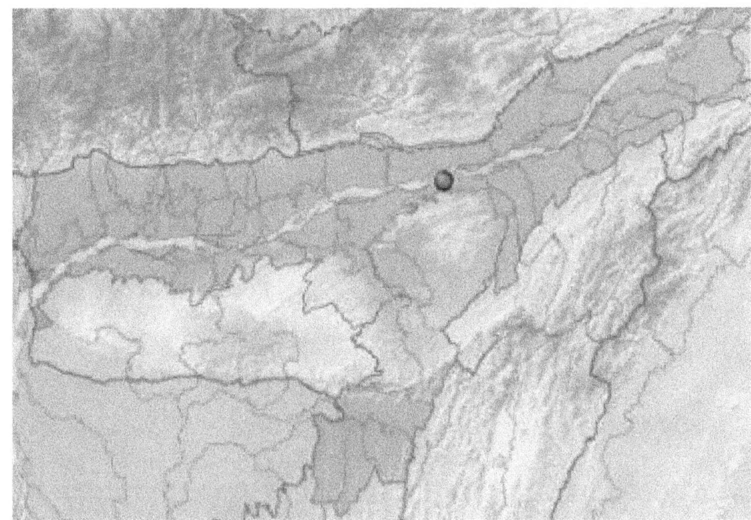

Figure: Location of Kaziranga National Park in Assam. Milenioscuro, CC BY-SA 3.0 <https://creativecommons.org/licenses/by-sa/3.0>, via Wikimedia Commons

15.5 Kaziranga National Park, Assam

Kaziranga is a UNESCO World Heritage Site and one of the most successful wildlife conservation stories in India. Home to two-thirds of the world's population of one-horned rhinoceroses, Kaziranga's floodplains and grasslands also support tigers, elephants, and numerous bird species. Strict anti-poaching measures and habitat protection have made this park a model for rhino conservation worldwide.

15.6 Sundarbans National Park, West Bengal

The Sundarbans are the largest tidal mangrove forest in the world, straddling India and Bangladesh. This UNESCO World Heritage Site is famed for its unique ecosystem, supporting a variety of wildlife, including the Royal Bengal tiger. The Sundarbans also provide critical services, acting as a buffer against cyclones and protecting the region's rich biodiversity.

15.7 Gir National Park, Gujarat

Gir National Park is the last stronghold of the Asiatic lion, a species once on the brink of extinction. Thanks to dedicated conservation efforts, the lion population has rebounded, making Gir one of India's most important protected areas. The park's dry deciduous forests also support leopards, hyenas, and numerous bird species.

15.8 Great Himalayan National Park, Himachal Pradesh

Located in the western Himalayas, the Great Himalayan National Park is a UNESCO World Heritage Site known for its stunning landscapes and high-altitude biodiversity. The park is home to species like the snow leopard, Himalayan brown bear, and Western tragopan, and is vital for the conservation of alpine ecosystems.

15.9 Wildlife Conservation Activities in Jim Corbett National Park

The activities for tiger conservation and other activities are described below.

Project Tiger: Jim Corbett is a flagship site for Project Tiger, launched in 1973 to conserve tigers and their habitats. The park's management focuses on habitat preservation, anti-poaching measures, and monitoring tiger populations. Regular tiger censuses using camera traps and tracking methods help in assessing population dynamics and ensuring the protection of this apex predator.

Anti-Poaching Patrols: To combat poaching, the park deploys anti-poaching units equipped with modern surveillance tools, such as drones and night-vision cameras. These patrols safeguard tigers from illegal hunting and trade.

Forest and Grassland Restoration: Conservation activities include managing and restoring forests and grasslands to maintain habitat quality. Controlled burns, invasive species management, and habitat enrichment activities ensure that the park's diverse ecosystems support a variety of species.

Water Resource Management: Maintaining water sources is crucial for wildlife health. The park's management ensures that water bodies are preserved and adequately maintained to support both terrestrial and aquatic wildlife.

Eco-Development Programs: Initiatives to involve local communities in conservation efforts include eco-development programs that provide alternative livelihoods and promote sustainable practices. Local communities are engaged in activities such as wildlife education, forest protection, and tourism management.

15.9 Wildlife Conservation Activities in Kaziranga National Park

The activities for rhino conservation and other activities are briefly described below.

Anti-Poaching Measures: Kaziranga has implemented rigorous anti-poaching measures to protect its rhino population. The park employs armed forest guards, uses surveillance technology, and maintains intelligence networks to combat poaching. Regular patrolling and community awareness campaigns help deter illegal activities.

Habitat Management: The park's management focuses on preserving and enhancing rhino habitats through activities such as controlling invasive plant species, managing grasslands, and restoring wetlands. These efforts ensure that the park's ecosystem can support a thriving rhino population.

Wildlife Surveys: Kaziranga conducts regular wildlife surveys to monitor the health and numbers of various species, including tigers, elephants, and birds. Camera traps, direct sightings, and aerial surveys provide data for effective management and conservation planning.

Research and Education: The park supports research on wildlife behavior, ecology, and conservation needs. Educational programs and outreach initiatives help raise awareness among visitors and local communities about the importance of preserving biodiversity.

Human-Wildlife Conflict Resolution: To address human-wildlife conflict, Kaziranga employs strategies such as building protective barriers, creating buffer zones, and engaging local communities in conflict management. Efforts to reduce incidents of crop damage and property destruction help maintain harmony between wildlife and people.

Community Participation: Involving local communities in conservation activities, such as anti-poaching patrols and habitat management, fosters a sense of ownership and collaboration in preserving the park's natural resources.

15.10. Conclusion

India's network of national parks, wildlife sanctuaries, and biosphere reserves forms the cornerstone of its biodiversity conservation efforts. These protected areas not only safeguard endangered species but also preserve entire ecosystems that provide critical ecological services.

Chapter 16: Climate Change and Its Impact on India's Wildlife

Climate change is one of the most pressing global challenges of our time, and its effects are particularly pronounced in biodiversity-rich countries like India. With its vast range of ecosystems—from the Himalayas to the Western Ghats, from arid deserts to lush mangroves—India's wildlife is deeply intertwined with its climate. This chapter explores how climate change is impacting India's wildlife, examining the consequences for species, habitats, and ecosystems, and highlighting ongoing and potential strategies for mitigating these impacts.

16.1 The Climate Change Landscape in India

India is experiencing significant changes in its climate due to global warming. Key climate impacts include:

Rising Temperatures: Average temperatures in India have been increasing, leading to more frequent and severe heatwaves.

Changing Precipitation Patterns: Shifts in monsoon patterns and increased variability in rainfall are altering the distribution and intensity of precipitation.

Extreme Weather Events: The frequency of extreme weather events, such as cyclones, floods, and droughts, has increased.

Sea-Level Rise: Rising global temperatures are causing polar ice melt and thermal expansion of seawater, leading to increased sea levels.

16.2 Impacts of climate change on India's wildlife: Habitat Loss and Alteration

Many of India's ecosystems are sensitive to temperature and precipitation changes. For example:

Himalayan Ecosystems: Glacial retreat in the Himalayas affects freshwater availability and alters habitats for species like the snow leopard and the Tibetan antelope. Changes in snow cover also impact the breeding and feeding patterns of these species.

Western Ghats: Rising temperatures are shifting the distribution of plant species in the Western Ghats, affecting the habitats of endemic species like the Nilgiri tahr and the Lion-tailed macaque.

16.3 Changes in Food Availability

Climate change impacts the availability of food sources for wildlife:

Marine Species: Ocean warming and acidification affect marine biodiversity, disrupting the food chains for species like the Indian mackerel and tuna. Coral reefs, which are critical for many marine species, are suffering from bleaching due to rising sea temperatures.

Forest Species: Changes in flowering and fruiting patterns of plants affect herbivores like the Indian elephant and gaur, as their food sources become less predictable.

16.4 Altered Migration Patterns

Climate change influences the migration patterns of many species:

Bird Migration: Changes in temperature and weather patterns affect migratory birds. For instance, the timing and routes of migratory species like the Amur falcon and Spoon-billed sandpiper are being disrupted, impacting their breeding success and survival.

Marine Turtles: Sea-level rise and increased temperatures impact nesting sites for marine turtles such as the Olive ridley turtle, leading to changes in nesting success and hatchling survival rates.

16.5 Increased Vulnerability to Diseases

Climate change can increase the prevalence of diseases affecting wildlife:

Vector-Borne Diseases: Warmer temperatures and changing precipitation patterns can expand the range of disease vectors like mosquitoes and ticks. This can lead to the spread of diseases such as anthrax and tick-borne fevers affecting wildlife populations.

Pathogen Proliferation: Higher temperatures can promote the growth of pathogens like fungi and bacteria, leading to increased disease outbreaks among wildlife.

16.6 Conservation Strategies and Adaptation Measures: Protecting and Restoring Habitats

Habitat protection and restoration are critical for helping wildlife adapt to climate change:

Protected Areas: Expanding and effectively managing protected areas can help safeguard critical habitats and create climate refuges for wildlife. The creation of wildlife corridors can facilitate species movement and adaptation.

Reforestation and Afforestation: Planting native trees and restoring degraded forests help to increase habitat resilience and support species that rely on specific forest conditions.

16.7 Monitoring and Research

Ongoing research and monitoring are essential for understanding climate change impacts and developing effective responses:

Biodiversity Monitoring: Regular monitoring of species populations and habitat conditions helps track changes and identify vulnerable species. This information is crucial for informing conservation strategies.

Climate Impact Research: Research into how climate change affects wildlife behavior, physiology, and distribution is important for developing targeted conservation measures.

16.8 Climate-Resilient Conservation Planning

Adapting conservation strategies to account for climate change is essential:

Adaptive Management: Conservation plans should incorporate flexibility to adapt to changing conditions. This may include adjusting protected area boundaries, managing for new threats, and incorporating climate projections into planning.

Ecosystem-Based Approaches: Focusing on ecosystem health and resilience can help support wildlife adaptation. For example, managing wetlands and riparian zones for water retention and flood control can benefit both wildlife and local communities.

16.9 Community Engagement and Education

Engaging local communities and raising awareness about climate change and its impacts on wildlife can support conservation efforts:

Community Involvement: Involving local communities in conservation initiatives, such as habitat restoration and wildlife monitoring, helps build support and ensure the sustainability of efforts.

Education and Advocacy: Educating the public about the importance of wildlife conservation and the impacts of climate change can foster a conservation ethic and encourage behavior changes that support wildlife protection.

16.10 Areas of focus for the future

Addressing the challenges posed by climate change requires a collaborative approach that integrates science, policy, and community action. Key areas for future focus include:

Enhancing Policy Frameworks: Developing and implementing policies that address climate change impacts on wildlife is crucial. This includes integrating climate considerations into wildlife protection laws, land-use planning, and conservation strategies.

Strengthening International Cooperation: Climate change is a global issue that requires international collaboration. Engaging in global climate agreements and conservation partnerships can enhance efforts to protect wildlife and their habitats.

Promoting Sustainable Development: Ensuring that development practices are sustainable and climate-resilient can reduce the impact of human activities on wildlife. This includes promoting green infrastructure, sustainable agriculture, and eco-friendly practices.

16.11 Conclusion

Climate change presents significant challenges to India's wildlife and biodiversity. The impacts of rising temperatures, shifting weather patterns, and extreme events threaten habitats, food sources, and species survival. However, through targeted conservation strategies, research, and community engagement, it is possible to mitigate these impacts and support wildlife adaptation.

Chapter 17: Role of Technology in the Conservation of India's Wildlife

Technology has revolutionized many aspects of our lives, and its impact on wildlife conservation in India is no exception. With a rich tapestry of biodiversity spanning lush forests, arid deserts, and sprawling wetlands, India faces the monumental task of protecting its wildlife amidst a backdrop of increasing human encroachment, climate change, and environmental degradation. Fortunately, technology offers innovative solutions to these challenges, enhancing the efficiency and effectiveness of conservation efforts.

Figure: A camera trap set up in the field, placed so that the beam is interrupted by the potential subject/animal to be captured. By User:Prashanthns - Own work, CC BY-SA 3.0, https://commons.wikimedia.org/w/index.php?curid=5623229

17.1 Advanced Monitoring and Data Collection

One of the most significant contributions of technology to wildlife conservation is the ability to monitor and collect data with unprecedented precision and scope.

Camera Traps: Remote camera traps, equipped with motion sensors, capture images of wildlife in their natural habitats without human disturbance. These devices provide crucial data on species distribution, population densities, and behavior. For instance, camera traps have been instrumental in monitoring the elusive Bengal tiger in the Sundarbans and the snow leopard in the Himalayas.

Satellite Imagery: Satellites offer a bird's-eye view of large landscapes, enabling conservationists to monitor deforestation, habitat changes, and illegal activities. High-resolution satellite images help track land use changes, forest cover loss, and the impact of natural disasters on wildlife habitats.

Drones: Drones equipped with cameras and sensors are used for aerial surveys, anti-poaching operations, and habitat assessments. They provide real-time data on wildlife movements and help in the rapid detection of illegal activities, such as poaching and logging.

17.2. Enhancing Anti-Poaching Efforts

Poaching remains a critical threat to many species in India. Technology plays a pivotal role in combating this menace through several innovative approaches.

Electronic Surveillance: Using GPS trackers and radio collars, conservationists can monitor the movements of endangered species, such as rhinos and tigers, in real-time. This technology not only helps in tracking animal movements but also alerts authorities to potential poaching threats.

Anti-Poaching Patrols: Advanced technologies, including night vision goggles and thermal imaging, assist anti-poaching patrols in detecting poachers and illegal activities even under challenging conditions. These tools improve the effectiveness of ground patrols and increase the chances of apprehending poachers.

17.3 Supporting Habitat Conservation

Protecting and restoring habitats is essential for wildlife conservation. Technology aids in these efforts through various means.

Geographic Information Systems (GIS): GIS technology helps in mapping and analyzing habitats, land use patterns, and conservation priorities. It supports the identification of critical habitats, planning of protected areas, and evaluation of conservation strategies.

Reforestation and Afforestation: Technologies such as drone-assisted seed planting and remote sensing are used to monitor and manage reforestation projects. These innovations improve the efficiency of habitat restoration efforts and ensure that newly planted areas are monitored for growth and health.

17.4 Engaging the Public and Raising Awareness

Technology also plays a crucial role in engaging the public and raising awareness about wildlife conservation.

Citizen Science: Mobile apps and online platforms enable the public to participate in wildlife monitoring and data collection.

Social Media and Digital Campaigns: Social media platforms and digital campaigns are powerful tools for spreading awareness about wildlife issues, promoting conservation initiatives, and mobilizing support. These platforms help in reaching a wide audience and generating public interest and involvement in conservation efforts.

17.5 Future Prospects and Challenges

As technology continues to advance, its role in wildlife conservation will likely expand. Emerging technologies, such as artificial intelligence (AI) and machine learning, have the potential to further enhance data analysis, predict wildlife trends, and improve conservation strategies.

However, the integration of technology in conservation also presents challenges. Ensuring that technology is accessible and affordable for all conservation practitioners, especially in remote areas, is crucial. Additionally, addressing concerns related to data privacy and the ethical use of technology in wildlife monitoring is essential.

17.6 Conclusion

Technology has become an indispensable ally in the conservation of India's wildlife. From advanced monitoring tools to anti-poaching innovations and public engagement platforms, technology enhances the capacity to protect and preserve the country's rich biodiversity. By embracing and advancing technological solutions, India can improve conservation of its wildlife in many areas.

Chapter 18: Conclusion: Conservation and the Future in India

India is home to some of the most diverse and iconic wildlife species in the world, ranging from the Bengal tiger and Indian elephant to the snow leopard and one-horned rhinoceros. The relationship between humans and wildlife in India is complex and has evolved over centuries. In recent times, with rapid development, population growth, and environmental challenges, the pressure on wildlife and natural habitats has increased. However, there are also strong conservation efforts to preserve India's biodiversity. This chapter explores the intricate relationship between humans and wildlife in India, examines current conservation efforts, and looks toward the future of wildlife conservation in the country.

18.1 The Historical Relationship Between Humans and Wildlife in India

Historically, humans and wildlife in India coexisted in relative harmony. Ancient Indian texts and traditions are filled with reverence for animals and nature. Animals such as tigers, elephants, and snakes hold important symbolic significance in Indian culture, appearing in religious myths, folklore, and art. The principles of living in harmony with nature, as articulated in ancient Hindu, Buddhist, and Jain texts, have long influenced Indian society.

18.2 Indigenous Practices

India's indigenous communities have traditionally played a key role in wildlife conservation. These communities have a deep understanding of ecosystems and biodiversity, often practicing sustainable hunting, fishing, and agriculture. The Bishnoi community of Rajasthan, for example, is famous for its protection of animals, particularly blackbucks, and trees, guided by their religious beliefs that emphasize the sanctity of all life.

However, as human populations grew, the demand for land, food, and resources increased, leading to habitat loss, hunting, and poaching, especially during the colonial period. Large game hunts, commercial exploitation of forests, and the introduction of industrial agriculture had devastating impacts on India's wildlife populations.

18.3 The Future of Wildlife Conservation in India

The future of wildlife conservation in India hinges on addressing both environmental and socio-economic challenges. The following are key areas where future efforts must be focused:

Integrating Conservation with Development: As India continues to develop, it is critical to integrate conservation goals with economic development. Sustainable development strategies must account for the protection of biodiversity and ecosystems, ensuring that infrastructure projects, such as roads and dams, minimize their impact on wildlife habitats.

Policies like eco-sensitive zones around protected areas and the promotion of green infrastructure can help mitigate the environmental impacts of development. Balancing the needs of growing human populations with the conservation of natural habitats will be a significant challenge moving forward.

Addressing Climate Change: Climate change adaptation and mitigation efforts are essential for the long-term survival of India's wildlife. Protecting and restoring ecosystems that act as carbon sinks, such as forests and wetlands, can help combat climate change while providing habitats for wildlife.

Conservationists must also develop strategies to help species adapt to changing environments, such as creating climate corridors that allow animals to migrate to more suitable habitats.

Expanding Protected Areas and Corridors: Expanding India's network of protected areas and establishing wildlife corridors are vital for the survival of wide-ranging species like tigers and elephants. Wildlife corridors, which connect fragmented habitats, allow animals to move freely between protected areas, reducing the risk of inbreeding and human-wildlife conflict.

In the Western Ghats, the creation of elephant corridors has helped reduce conflicts between elephants and farmers while maintaining genetic diversity among elephant populations.

Empowering Local Communities: Empowering local communities to take ownership of conservation efforts is crucial. Providing economic incentives, such as eco-tourism and sustainable agriculture, can encourage communities to protect wildlife and their habitats. Involving indigenous knowledge in conservation strategies will also enhance the effectiveness of these efforts.

Strengthening Enforcement and Reducing Wildlife Crime: Strengthening law enforcement and addressing illegal wildlife trade must remain a top priority. Improved surveillance technologies, such as camera traps and drones, along with increased patrolling in protected areas, can help deter poaching. International cooperation is also essential in combating the transnational illegal wildlife trade.

18.4 Conclusion

The relationship between humans and wildlife in India is at a critical juncture. While the country faces immense challenges due to habitat loss, human-wildlife conflict, poaching, and climate change, there are also significant conservation successes and a robust framework for the future. By integrating conservation with development, empowering communities, and addressing emerging challenges like climate change, India can ensure a future where both humans and wildlife thrive in harmony.

About the author

113

Siva Prasad Bose is an author of introductory guidebooks on aspects of Indian laws. He is currently retired after many years of service as an electrical engineer in Uttar Pradesh Power Corporation Limited. He received his engineering degree from Jadavpur University, Kolkata and has a law degree from Meerut University, Meerut and a BSc from MMH College, Ghaziabad. His interests lie in the fields of family law, environmental law, law of contracts, and areas of law related to power electricity related issues. He lives in Delhi.

Other Books by Siva Prasad Bose

Introduction to Wills and Probate

Senior Citizens Abuse in India

Introduction to Negotiable Instruments

Introduction to Marriage Laws in India

Neighbor Problems in India and what to do about them

Delays in Court Cases in India

Self-Publish Books and E-Books in India

Introduction to Patents and Patent Law in India

Introduction to Ecology and Environmental Laws in India